T0186146

CURRENT ISSUES IN COSMOLOGY

What are the current ideas describing the large scale structure of the Universe? How do they relate to the observed facts? This book looks at both the strengths and weaknesses of the current big-bang model in explaining certain puzzling data. It arises from an international conference that brought together many of the world's leading players in cosmology. In addition to presenting individual talks, the proceedings of the resulting discussions are also recorded. Giving a comprehensive coverage of the expanding field of cosmology, this text will be valuable for graduate students and researchers in cosmology and theoretical astrophysics.

JEAN-CLAUDE PECKER has had a long and successful career of research in the theory of stellar atmospheres, and later, cosmology. After studying at the École Normale Supérieure and gaining his doctorate from Institut d'Astrophysique de Paris in 1950, he held research positions at the CNRS, the University of Clermont-Ferrand, and the Paris Observatory in Meudon. He was Director of the Nice Observatory (1962–69), General Secretary of the IAU (1964–67), Director of the Institut d'Astrophysique de Paris (1972–79), and Professeur at the Collège de France from 1963 to his retirement in 1988. He has devoted much of his time to UNESCO activities, to the defence of human rights, to the promotion of a rationalist view of the world, and to the popularization of science. He has written nearly 30 books on astronomy and solar astrophysics, many of which have been translated and distributed throughout the world.

Professor Pecker is Commandeur of the Légion d'Honneur, Grand'Croix of the Ordre National du Mérite, and Commandeur of the Palmes Académiques. He has received several international prizes and is a fellow of several academies, notably the Académie Royale de Belgique, the International Academy of Humanism, and the Academia Europaea. He is an Associate of the Royal Astronomical Society, London.

JAYANT V. NARLIKAR is a well-known cosmologist, noted for his opposition to the paradigm of the big bang and his work on alternative cosmologies. His early education was at the Banaras Hindu University, India, followed by higher degrees at the University of Cambridge (B.A. 1960, Ph.D. 1963, Sc.D. 1976). He was a fellow at King's College, Cambridge (1963–72) and a founder staff member of the

Institute of Theoretical Astronomy at the University of Cambridge (1966–72). He then returned to India to take up a professorship at the Tata Institute of Fundamental Research, where he led the Theoretical Astrophysics Group until 1989. In 1988 he was invited to set up the Inter-University Centre for Astronomy and Astrophysics, of which he was the Founding Director until his retirement in 2003. In addition to his research and travel, he has played a role as a popularizer of science, and *Brahmand*, his serial on astronomy in Hindi, enjoyed popular acclaim on Indian television. He has written many books, including *Introduction to Cosmology* (1993, 2002), The *Lighter Side of Gravity* (1996), and *Seven Wonders of the Cosmos* (1999).

Professor Narlikar has received several awards including the S.S. Bhatnagar Award for the Physical Sciences, The Janssen Medal of the Société Astronomique de France, and UNESCO's Kalinga Award for science popularization. He is an Associate of the Royal Astronomical Society, London, Fellow of the Third World Academy of Sciences, and a fellow of the three national academies of science in India.

CURRENT ISSUES IN COSMOLOGY

Edited by

JEAN-CLAUDE PECKER

Formerly at the Collège de France, Paris

and

JAYANT NARLIKAR

Formerly at the Inter-University Centre for Astronomy and Astrophysics, Pune, India

CAMBRIDGE
UNIVERSITY PRESS

CAMBRIDGE UNIVERSITY PRESS
Cambridge, New York, Melbourne, Madrid, Cape Town,
Singapore, São Paulo, Delhi, Tokyo, Mexico City

Cambridge University Press
The Edinburgh Building, Cambridge CB2 8RU, UK

Published in the United States of America by Cambridge University Press, New York

www.cambridge.org
Information on this title: www.cambridge.org/9781107403437

First published 2006
First paperback edition 2011

A catalogue record for this publication is available from the British Library

ISBN 978-0-521-85898-4 Hardback
ISBN 978-1-107-40343-7 Paperback

Contents

Contributors

Halton ARP, MPI, Garching, Germany

Francis BERNARDEAU, CEA/Saclay, France

Francesco BERTOLA, University of Padova, Italy

Christian BIZOUARD, Observatoire de Paris, France

Alain BLANCHARD, Observatoire Midi-Pyrénées, Toulouse, France

François BOUCHET, IAP, Paris, France

Henrik BROBERG, Stockholm, Sweden

Georges BUCHNER, Paris, France

Geoffrey BURBIDGE, UCSD, La Jolla, California, USA

Margaret BURBIDGE, UCSD, La Jolla, California, USA

Brandon CARTER, Observatoire de Paris-Meudon, France

Michel CASSÉ, CEA/Saclay, France

Michel DECOMBAS, Paris, France

Michael J. DISNEY, University of Wales, Cardiff, UK

Florence DURRET, IAP, Paris, France

Hédi EL AGREBI, Paris, France

Roger FERLET, IAP, Paris, France

Marcel FROISSART, Collège de France, Paris, France

Pasquale GALIANI, Italy

Francisco GOMEZ-MONT, Mexico City, Mexico

Georges GONZALES, SAF, Paris, France

Bruno GUIDERDONI, IAP, Paris, France

Carlos GUTTERIEZ, IAC, Tenerife, Canary Islands, Spain

Hayk HARUTYUNIA N, Bjurakan Observatory, Armenia

Georges HOYNANT, SAF, Paris, France

Roy C. KEYS, APEIRON, Montréal, Canada

Ralph KRIKORIAN, IAP & Collège de France, Paris, France

Valérie de LAPPARENT-GURRIET, IAP, Paris, France

Bernard LEMPEL, SAF, Paris, France
Christian MARCHAL, ONERA, Paris, France
Pascal MASSÉ, Paris, France
Roya MOHAYAEE, OCA, Nice, France
Mariano MOLES, Instituto de Astrofisila de Andalucía, Granada, Spain
Jacques MORET-BAILLY, Université de Bourgogne, Dijon, France
William NAPIER, Armagh Observatory, Northern Ireland, UK
Jayant V. NARLIKAR, Collège de France, Paris, France, & IUCAA, Pune, India
X. NEDJAR, Paris, France
Keith OLIVE, University of Minnesota, Minneapolis, USA
Georges PATUREL, Observatoire de Lyon, Saint-Genis-Laval, France
Jean-Claude PECKER, Collège de France, Paris, France
Huguette QUINTIN, Paris, France
Claude PICARD, SAF, Boulogne-Billancourt, France
Roland QUERRY, SAF, Paris, France
David ROSCOE, Sheffield Observatory, UK
Francis SANCHEZ, Université Paris XI, France
Claude SAROCHI, Paris, France
Manuel SMIL, Paris, France
Jack SULENTIC, University of Alabama, Tuscaloosa, USA
Jacques SUPERNANT, Paris, France
Jean SURDEJ, Université de Liège, Belgique
Elizabeth VANGIONI-FLAM, IAP, Paris, France
Kameshwar C. WALI, Syracuse University, NY, USA
N. Chandra WICKRAMASINGHE, Cardiff Centre of Astrobiology, Wales, UK
Heiner WIZEMANN, Paris, France

Preface

The idea of a Colloquium on ***"Cosmology: Facts and Problems"*** was mooted when one of us (JVN) was to visit Collège de France as Professor (Chaire Internationale) during 2003–04. Both of us felt that the subject of cosmology has seen considerable advancement on both observational and theoretical fronts but that there are many issues of observational nature that will remain to be understood. With this point of view the Colloquium was arranged during June 8–11, 2004, at Collège de France.

The Colloquium attracted leading workers in the field. They could be divided into three categories: 1. Observers 2. Theoreticians who liked to explain all the observed data in terms of the standard big-bang paradigm 3. Theoreticians who felt that there were some observations that did not allow a standard interpretation. Sometimes the observers also fell under categories 2 and 3. We were happy that the Colloquium attracted good participation from several countries and there was amiable and frank discussion on various issues. We had allowed plenty of time for discussion after each presentation including a panel discussion at the end. The proceedings presented here reflect this openness of the debate. Several participants who had not given a formal presentation also took part in the discussion.

We would like to express our grateful thanks to all those who helped us in various ways towards making this Colloquium such a success. In particular, we would like to thank Professor Jacques Glowinski, Administrateur du Collège de France, for his kindness in hosting the Colloquium at the Collège de France. We would also like to express our appreciation to Dr Ralph Krikorian, Maître de Conferences au Collège de France, Mme Simone Lantz, M. Jean-Claude Couillard, and Yvan Le Bozec from the Chaire de Physique Corpusculaire et Cosmologie du Collège de France for their help during the conference, and the Foundation Hugot and its Director, Mme Florence Terrasse-Riou, for financial support for the meeting. The administrative and technical staff of Collège de France also helped us in many different ways. In Pune, we thank Mr Vyankatesh Samak for his invaluable help in putting the manuscript together. Finally, we thank Dr Simon Mitton and Cambridge University Press their help in publishing these proceedings.

Jean-Claude Pecker
Jayant V. Narlikar

Part I

Observational facts relating to discrete sources

1

The state of cosmology

Geoffrey Burbidge

University of California, San Diego, La Jolla, CA, USA

In introducing the general topic of this meeting I am going to give a personal view. Only late in my professional career (~1990) did I begin to work seriously in cosmology, though I had always followed with interest the various claims that progress was being made, and I even wrote a review of the state of affairs for *Nature* in 1971 entitled, "Was There Really a Big Bang?" (Burbidge 1971).

1 Introduction

For some years this period, starting in the 1990s, has been said to be the golden age of cosmology. Compared with the situation earlier, this is a fair judgement, since in the last decade or more there has been a tremendous increase in the number of people working in the field, and large sums of money have been invested in new methods of observation of the background radiation and of large numbers of galaxies and other discrete objects, often those with high redshifts. Another important ingredient is the renewed interest in cosmology taken by many theoretical physicists and experimental particle physicists.

With this expansion has come a great deal of new information, and a model for the Universe that almost everyone believes in. This in turn means that while there are many conferences on cosmology, the theme is almost always the same. This meeting will be different because some of its organizers have for a variety of reasons not followed the main stream. At the same time I hope that there will be a fair discussion of the conventional cosmological model.

In this introduction I want to make it clear why it is that some of us do not accept as the only starting point the usual model of an evolving universe starting with an initial creation process. The arguments against this approach are of two kinds. First there is the history, which shows that on several occasions in the early work assumptions were made that would lead to the observed answers, when alternatives were possible i.e., there have been very few real predictions, and second, the modern

3

situation in which not all of the data are taken into account. This being the case it is extravagant and entirely premature to make the kind of claims that are now being made (cf. Spergel *et al.* 2003) for a standard model.

2 The expansion of the Universe

The major discovery was the redshift–apparent magnitude relation for nearby galaxies by Hubble in 1929 (Hubble 1929). This was immediately interpreted as direct observational evidence for an expanding universe of the Lemaitre–Friedmann type, meaning that this interpretation agrees with the expanding solution of Einstein's equations. By 1930 everyone accepted that the Universe is expanding. Reversal of the time axis of the expansion then leads to the conclusion that there was a finite origin for the Universe, which Lemaitre in 1936 originally described as the "Primeval Atom."

3 Nucleosynthesis and the cosmic microwave background

There were no convincing physical investigations of the early state of this Primeval Atom until the late 1940s, when a group of leading physicists including Rudolf Peierls, Enrico Fermi, Edward Teller, Maria Meyer, George Gamow, and his colleagues Ralph Alpher and Robert Herman and others made the assumption that it was at that very early epoch that the chemical elements were synthesized. Gamow in 1946 had originally speculated that the electron degeneracy in the early Universe would more than compensate for the mass difference between the neutron and a proton plus electron. Thus he concluded that the matter at the beginning would be a single neutron lump, so that the synthesis of the chemical elements out of this lump could be a verification of the Friedmann model. However, the problems of nucleosynthesis immediately encountered were, first, that there is no stable mass at $A = 5$ or 8 so that the build-up cannot go beyond D, ^{3}He, ^{4}He, and ^{7}Li. Second, a radiation field together with neutrons, protons, and electrons leads to more complications, which were discussed by Gamow, Alpher, and Herman. The other leading physicists gave up the problem when they realized that the bulk of the chemical elements could not be made in this way.

It was also realized in this period that the bulk of the known ^{4}He, approximately 25–30% by mass, could not have been made in the stars seen in the galaxies. The problem was that using the known luminosities of galaxies and the time scale for the Universe, which was then thought to be 2×10^9 years, very little helium would have been made. Thus it was concluded that the helium must have originated in primordial nucleosynthesis. This required that the energy density of radiation in the early Universe had to be very large. Until then, the reverse had always been assumed in Friedmann models. In such models S(t) (the scale factor) $\propto t^{1/2}$ and $T_9 = \text{const.} \, t^{-1/2}$. The next step was completely ad hoc. The mass density of stable non-relativistic particles, ρ_b, explicitly neutrons and protons in the theory of 1950, decreases with the expansion as S^{-3} or $t^{-3/2}$. Alpher and Herman put the density $\rho_b = 1.70 \times 10^{-2} \, t^{-3/2} \, \text{gm}^{-3}$. But there is nothing in the theory that fixes the value of this numerical coefficient. It is adopted to make things come out right, i.e., to make the calculated value of Y agree with the observed value. This is why the big-bang theory cannot be claimed to explain the microwave background or to explain a cosmic helium value close to 0.25. It is only an axiom of modern big-bang cosmology, and the supposed explanation of the microwave background is a restatement of that axiom. Thus in no sense did the big-bang theory predict the microwave background.[1] This would only be true if the factor 1.7×10^{-2} is called a prediction. If we eliminate t between the relations given above we find that

$$\rho_b = 1.51 \times 10^{-32} \, T^3 \tag{1}$$

which can be rounded off to $\rho_b \approx 10^{-32} \, T^3 \, \text{gm cm}^{-3}$. Alpher and Herman put the mass density of the Universe as $\rho_b = 10^{-30} \text{gm cm}^{-3}$ and thus concluded that T must be about 5 K. Ten years later, when the Hubble constant had been further reduced, it appeared that $\rho_b \simeq 10^{-29} \text{gm cm}^{-3}$, and then both Gamow and Dicke suggested that $T \simeq 15 \, \text{K}$. Of course these were gross overestimates.

What none of the physicists throughout this period was aware of was that in 1941 McKellar (1941) had determined the temperature of the interstellar radiation from the spectra of the interstellar lines due to the molecules CH and CH^+, which Adams and later McKellar had detected in the spectra of stars. McKellar showed that if the radiation has black-body form, 1.8 K $< T <$ 3.4 K, which is in remarkable agreement with what was found later. McKellar stated the following

"Adams has kindly communicated to the writer his estimate of the relative intensity, in the spectrum of ξ Ophiuchi, of the λR(0) interstellar line of the λ3883 CN band and the λ3874.00, R(1) line, as 5 to 1. $B_0J''(J'' + 1) + \ldots$ has the value 0 and 3.78 cm^{-1} for the 0 and 1 rotational states and for the two lines R(0) and R(1) the values of the intensity factor i are, respectively, 2 and 4. Thus from (3) we find, for the region of space where the CN absorption takes place, the "rotational" temperature,

$$T = 2.3 \, K.$$

If the estimate of the intensity of R(0)/R(1) were off by 100 per cent, this value of the "rotational" temperature would not be changed greatly, R(0)/R(1) = 2.5, giving T = 3.4 K and R(0)/R(1) = 10 giving T = 1.8 K."

---

[1] For Y = 0.24, which is closer to the preferred current value, the constant η is now close to 4.5×10^{-2}.

When, in 1965, Penzias and Wilson reported that they had directly detected the radiation (Penzias and Wilson 1965) and later Mather *et al.* (1990, 1994) showed that the radiation is of almost perfect black-body form with $T = 2.726$ K, they were richly rewarded. What I want to stress here is that while the black-body nature of the radiation was predicted by the big-bang theory, the numerical value of the temperature was not, and cannot be (see Turner 1993), and since McKellar had already measured it, admittedly indirectly, it is a moot point as to whether the 1965 event truly was a major discovery. (If someone has already discovered a new phenomenon and published it, but the people most interested are unaware of the earlier discovery, how should credit be apportioned?) In truth no prediction was involved. But the psychological effect based on mistaken ideas concerning the prediction and discovery is one of the major reasons why the big bang theory is believed.

What is now being done is to put the observed temperature in Equation (1) and derive a value for ρ_b. This is then compared with the value obtained from the nucleosynthesis calculations and observations involving D, ^{3}He, and ^{4}He. Very good agreements can be reached between theory and observation for $\rho \sim 3 \times 10^{-31}$ gm cm^{-3}; so this is now called the observed baryonic mass fraction in the Universe. This is a clear plus for the big-bang cosmology. However, since the closure density in the big-bang model $3H_0^2/8\pi G$ is about 6.8×10^{-30} gm cm^{-3} (for $H_o = 60$ km s^{-1} Mpc^{-1}) this is only about 5% of the closure density.

While this discrepancy has been known for $\sim$30 years, it is only in the last few years that this "missing" mass energy has been claimed first to be cold dark matter (CDM) and more recently cold dark matter and dark energy (Λ CDM).

An elaborate "theory" (more appropriately a "scenario") of galaxy formation then rests on this belief that this missing mass is real, because only if CDM exists in large measure is it possible to simulate galaxy formation at all. This is a classical example of "The Emperor has no clothes" syndrome. While a great deal of energy and money is being devoted by particle physicists to searches for the WIMPS, which could conceivably be the basis for the dark matter, nothing has been found so far (cf. Seife, 2004).

But, of course, none of this is necessary if we go back to the original observation of the ^{4}He/H ratio and take the position that the observed ratio is the result of hydrogen burning in stars. Then, of course, the whole of the mass must be baryonic. This leads us to one final point. If hydrogen burning was responsible for this ratio, an estimate can be made directly from observation of the energy released in this process. The mass density in the Universe can be determined from the masses of galaxies derived from their rotation curves and/or the velocity dispersion of the stars in galaxies, or of the galaxies in clusters. The virial for both individual galaxies and clusters is assumed to hold, so that in making this estimate we are assuming that some of the mass is dark. Putting in observed values for the space density of galaxies and a range of values of *M/L*, and a Hubble constant of 60 km s^{-1} Mpc^{-1} and supposing that the ^{4}He/H ratio is 0.24, we obtain an energy density of the radiation 4.5×10^{-13} erg cm^{-3}. This energy will initially be released in hard photons (UV radiation) but ultimately, according to thermodynamic arguments, it will be degraded to black-body radiation with $T \simeq 2.75$ K. This is remarkably close to the measured value of 2.726 K. This is either a pure coincidence, as it must be for those who believe in the big bang, or else it tells us that hydrogen burning was originally responsible for the Cosmic Microwave Background (CMB). In the Quasi-Steady-State Cosmology (QSSC) it is argued that it is due to hydrogen burning in the newly created galaxies and that intergalactic dust is responsible for the degradation to thermal energy.

While this agreement was mentioned in one or two earlier papers (cf. Fowler, Wagoner, and Hoyle 1967) it was not described in detail until 1998 when Hoyle and I managed to get it published in the *Astrophysi. J.* (Burbidge and Hoyle 1998). The paper was earlier rejected by *Phys. Rev. Lett.*, whose referees were strong proponents of the big bang. In our paper we showed that it was possible to explain the origin of all of the isotopes including D and ^{3}He in stars. D is probably built up in stellar flares on the surfaces of stars and partly destroyed by mixing in stellar interiors. An observational fact following from this hypothesis is that it predicts the D/H will be variable from one place in the galaxy to another, from galaxy to galaxy, and from QSO to QSO. But there really is no need to invoke a big bang.

Since none of the observations just described require this, what are the alternatives? Since the universe *is* expanding we can consider as possibilities a steady-state universe, which remains unchanged, or a cyclic universe with a cycle period of ∼20 Gyr. Here we omit discussion of Milne's kinematic cosmology, though it should not be forgotten that Milne raised the problem of the particle horizon, in the classical big-bang picture, and this is only claimed to be resolved now by recourse to an inflationary period.

It is natural that what came next was the classical steady-state universe of Bondi and Gold (1948) and Hoyle (1948).

4 The steady-state universe

The basic idea is that the Universe is not evolving. Thus matter (hydrogen) must be spontaneously created at a rate determined by the expansion. Bondi and Gold (1948) used as the basis for the theory what they called the perfect cosmological principle. Hoyle (1948) obtained the same model by generalization of Einstein's theory allowing for a repulsive term in the strong field regime (the C field) corresponding to creation (cf. Hoyle and Narlikar 1964, 1966). The steady-state theory was given quite a hostile reception as can be seen from an appraisal of the ways in which the various observational tests of the theory were handled (cf. Dingle 1953; Hoyle 1969; Hoyle *et al.* 2000, Chapter 7).

I believe that much of the prejudice in modern cosmology began at this time. In general the observers did not like the steady-state theory, although several of the pieces of observational evidence against it were shown later to be false. My good friend Allan Sandage has always insisted that some of his colleagues at Mount Wilson and Palomar were from its inception convinced that the steady state must be wrong, because they already had good evidence for evolution. Overall, one has the impression that most people liked the idea that there was a beginning, and that evidence for evolution would ultimately be detected. The general view was that all of the galaxies are old with ages comparable to H_0^{-1}. Thus, for example, evidence for young galaxies with ages $\ll H_0^{-1}$ (cf. Burbidge, Burbidge, and Hoyle 1963) was immediately disputed (Sandage 1963), so fast indeed, that the rebuttal paper of Sandage was published ahead of the paper by Burbidge *et al.* (was the editor, a good friend of all of us, showing his prejudice?).

5 The acceleration

There was one clear-cut prediction from the steady-state theory. This was that the expansion of the Universe would tend to accelerate (due to the creation process) rather than decelerate, as it must do in all Friedmann models without a cosmological constant (cf. Hoyle and Sandage 1956). Thus many claims were made from 1950 onwards that the observations showed that the Universe is decelerating, until by the 1980s it was finally admitted that the uncertainties in the observational methods being used were so great that it was impossible to decide.

Much more recently, starting in 1998, work using supernovae of Type Ia as standard candles, which can be detected at high redshifts, was announced by Perlmutter, Riess, and their colleagues (Perlmutter *et al.* 1999; Riess *et al.* 1998). They showed fairly conclusively, initially, with measurements out to $z \simeq 0.6$ that the Universe *is* accelerating. This being the case, there are two different cosmological scenarios that can explain it. The first is to insert a positive cosmological constant into the usual Friedmann models. The second is to remember that the classical steady-state theory *predicted* (cf. Hoyle and Sandage 1956) this result and the modified steady state (the QSSC) also predicted that the Universe would be accelerating (Hoyle *et al.* 1993, 2000). However, in reporting this result the observers once again showed their prejudice. Instead of at least stating that their result was qualitatively what had been predicted by the classical steady-state model and the quasi-steady-state cosmology, as is normally done in announcing a new observational result, and then going on to interpret their data in terms of a Friedmann model with a positive cosmological constant, they simply made the claim that they had demonstrated the reality of that model, as though that was the only way to go. And, of course, in doing this they were followed by the community who were equally ignorant or biased, or both, though attempts to clarify the situation (cf. Narlikar *et al.* 2002) have been published.

6 Driven by the cosmic microwave background (CMB) and the NASA value of H_0

Since the direct discovery of photons from the CMB by Penzias and Wilson in 1965, and the mistaken belief by many that this was the fulfillment of a prediction by Gamow and his colleagues (though they were undoubtedly short-changed when it came to recognition), the standard model largely buttressed by this CMB "discovery" took over. It was generally *assumed*[2] before it was established that the radiation would have black-body form (cf. the continuous discussion of "*relict*" radiation by the school of Zel'dovich), as indeed had been predicted by Gamow *et al.*, provided it was generated in the big bang, and when it was finally showed by Mather *et al.* (1990, 1994) that the radiation has a beautiful black-body form over a wide range of wavelengths the triumph was complete. The result was cheered at the meeting when it was first announced (I was the chairman of the session of the AAS meeting at which the announcement was made).

For nearly all cosmologists this was thought to be the death knell of the steady-state model and any of its improvements (which we were working on at the time). The idea that such a background spectrum could be obtained from many discrete sources appeared to be much too farfetched, though we have now shown that it is entirely possible (Hoyle *et al.* 2000). And in many ways what was more important, the CMB had shown how homogeneous and isotropic this component of the Universe is. But a serious question that was still unanswered was to understand how the matter component can also show the same effect on the large scale, i.e., homogeneity and isotropy, if galaxies first condensed from quantum fluctuations in a very early universe, when conditions prevailed such that objects were not able to communicate with each other soon after the beginning.

The way out of this problem was to invoke *inflation*, proposed by Guth (1981) and Linde (1982, 1983). The main point that I want to make here is not that inflation is not a good idea. It *is*, but it is *not a paradigm* (cf. Peebles 1993). It is yet another idea *invented* to explain what we see, like the numerical value of the initial baryon-to-photon ratio and the existence of non-baryonic matter. Inflation has no basis in fundamental theory. Given all three of these assumptions we can make a plausible

---

[2] Preliminary observations from rockets suggesting that the background radiation was not of black-body form were widely discredited by theorists who had already made up their minds.

model that will fit the observations. *Without them we cannot.* But this is how big-bang cosmology or, if you like, evolutionary cosmology has progressed. The most recent observational programs are devoted to fitting together more and more of the details based on a series of assumptions chosen to make the original model work.

Undoubtedly the most impressive work of late on models of the Universe has been the most recent analysis of the CMB based on the WMAP observations. Spergel *et al.* (2003) have shown that assuming a model in which the Universe is flat with a large cosmological constant Λ, in which galaxy formation was started by nearly scale invariant adiabatic Gaussian fluctuations, they can fit the WMAP data very well with other parameters such as the Hubble constant and the D/H ratio in high redshift QSOs.

The agreement between the model calculations of the acoustic fluctuations in the CMB due to matter fluctuations out to the third peak expected is particularly impressive, so that there now is considerable interest and belief in this latest "cosmological concordance" model.

However, if we restrict ourselves to observational quantities that are not based on any assumptions other than that the Universe is expanding, the greatest discrepancy between model parameters chosen and observations probably comes from the Hubble constant, which Spergel *et al.* have used. They have claimed that this best-fit model is obtained when $H_o = 71 \, \mathrm{km \, s^{-1} \, Mpc^{-1}}$, almost exactly the same as the value claimed to be correct by the group working with the Hubble Space Telescope (HST) and called the HST Key Project (Freedman *et al.* 2001). The difficulty with this is that this value of H_o may be much too high. Sandage and Tammann, the most experienced workers in the field, have since 1974 argued that a value close to $50 \, \mathrm{km \, s^{-1}}$, $\mathrm{Mpc^{-1}}$ is a much better choice (for a detailed discussion see Hoyle *et al.* 2000, Chapter 4). Over the last few years Sandage and Tammann have competed directly with the other group, also using the HST (Tammann *et al.* 2002), but for reasons much more to do with NASA's approach to public relations than to science, all of the publicity and attention has been given to the results and the personalities of Freedman *et al.* When we made a careful study of all of the data available up to 1999 (Hoyle *et al.* 2000) we concluded that the best value is $H_o = 56$–$58 \, \mathrm{km \, s^{-1} \, Mpc^{-1}}$. Sandage and Tammann and their colleagues in their most recent work (Parodi *et al.* 2000 Tammann; et al., 2002) have obtained a value for $H_o = 58.5 \, \mathrm{km \, s^{-1} \, Mpc^{-1}}$.

There is no doubt that the popularity of the higher value of H_o has much more to do with the sociology of astronomy than to science. In this case the origin of this belief can be dated rather precisely, to May 25, 1999, when NASA held a press conference in Washington to announce, as they modestly put it, that the search for the Holy Grail of cosmology was over.[3] The research team working on what was called the Hubble Space Telescope Key Project claimed they had finally solved one of the

---

[3] (Ref. *New York Times Magazine* July 25, 1999). The article is entitled "The Loneliness of the Long-Distance Cosmologist," and it is not a very nice article about Allan Sandage.

original mysteries – the age of the Universe. Sandage and his team were barred from attending or speaking at this press conference. The press conference was followed up by a similar announcement to a very large group at an AAS meeting in Chicago, this time by Robert Kennicutt, a Key team leader. He was more circumspect and mentioned Sandage's and Tammann's work. Kennicutt's announcement was also widely publicized, as was the work on the microwave background mentioned earlier.

How sensitive is the model to the value of H_0 that is put into the calculations?

Recently Blanchard *et al.* (2003) have tested whether or not a cosmological constant is really required by these observations of the CMB and large scale structure. They find that it is not, provided that the value of the Hubble constant is 46 km s^{-1} Mpc^{-1}, a value that is certainly compatible with the work of Sandage and Tammann. Other quite small changes in other parameters are required. Then we are back to an Einstein–de Sitter model. But then we have to deal with the evidence for acceleration described earlier, because it was this evidence that led the community to do an about turn soon after about 1998 and start using a positive cosmological constant.

Five years after the first evidence for acceleration and hence the presence of a positive cosmological constant "dark energy" was claimed, the picture has become more complicated. Many more SN Ia redshifts have been obtained out to redshifts $z \sim 1.5$ (cf. Barris *et al.* 2004; Riess *et al.* 2004). While it is still claimed that the work shows that there is dark energy and dark matter, it is suggested that at a redshift of about 0.5 there was a transition between acceleration and deceleration (a cosmic jerk). It also appears that a model with no cosmological constant, in which the effect is due to dust that is replenished at the same rate as it is diluted by the expansion, could also explain the observations (Riess *et al.* 2004).

By showing the way that a standard model has evolved (always starting with a big bang), I hope that by now that I have provided enough evidence for a reasonable person to conclude that there is no particularly compelling reason why one should so strongly favor a standard model Universe starting with a beginning rather than an alternative approach, apart from the fact that it is always easier to agree with the majority rather than to disagree. This sociological effect turns out to be actually extremely powerful in practice, because as time has gone on young cosmologists have found that if they maintain the status quo they stand a much better chance of getting financial support, observational facilities, and academic positions, and can get their (unobjectionable) papers published.

7 Explosive phenomena and the alternative cosmological approach

Starting in the 1950s the first radio galaxies were identified, and it became clear that they are extremely powerful energy sources often emitting energies of at least 10^{60} ergs ($\simeq 10^6$ M$_\odot$) in the form of relativistic particles and magnetic flux filling large volumes outside the galaxies, though they must have arisen from very small nuclei. The origin of this energy is either gravitational or is due to creation in galactic nuclei. It soon became clear that many galaxies have active energetic nuclei, emitting large fluxes of energy not only in radio frequencies but in optical, X-ray, and γ-ray wavelengths. Thus, by the 1980s it was generally accepted that explosive events in galaxies are of primary importance.

But, from the point of view of cosmogony, this comparatively new phenomenon has not been integrated into the classical evolutionary cosmological picture.

However, in the 1950s and 1960s V. A. Ambartsumian (1958, 1961, 1965) had already made the radical proposal that the centers of galaxies are places where the material of new galaxies is created and ejected. While Ambartsumian's ideas, based completely on observations, have been largely ignored by the cosmological establishment, these are the cosmogonical ideas out of which, in the 1990s, Hoyle, Narlikar, and I formulated the quasi-steady-state cosmology (QSSC) in which it is argued that the centers of active galaxies *are* the creation sources, and it is in them, in the vicinity of near black holes, that the C (Creation)-field operates. Thus matter is being created out of a set of singular points associated with the nuclei of galaxies. Thus, using biblical terminology, galaxies do beget galaxies. This leads to expansion and contraction with a period of about 40×10^9 years superimposed on an overall expansion with a characteristic time $\sim 10^{12}$ years. This is a cyclic universe, which does not contract to extremely small dimensions (Narlikar and Burbidge 2004).

This theory, based on the C-field theory of Hoyle and Narlikar (1964, 1966), was developed in a number of detailed papers published since 1990 and in a book (Hoyle *et al.* 2000). Although many details remain to be worked out, it seems possible that all of the observed properties of the Universe can be understood within the framework of this theory, though there are some phenomena that are still extremely difficult to understand. At this meeting these data will be discussed in detail by several speakers and Dr. Narlikar will discuss the QSSC.

8 Conclusion

In this introductory talk I have tried to describe some of the historical evidence that suggests to some of us that in trying to understand the properties of the Universe we should not be forced into the straight jacket of the standard cosmology.

It is certainly possible that everything did begin in a single explosion, or, if the Universe is cyclic (Steinhardt and Turok 2002, for example), it collapses to very small dimensions before it bounces, but this may not have ever happened. If it did, theoretical physicists may relate it ultimately to string theory or something related to it, but this will still for ever be out of reach of any genuine observational test.

If, on the other hand, ejection is occurring in the centers of galaxies all around us and throughout the Universe, we stand a much better chance of making observations that we may ultimately be able to interpret in theoretical terms.

Partly for historical reasons, and partly because cosmological research is driven more by the beliefs of strong individuals than by observational evidence, there is a complete imbalance between very different, but viable approaches. Only one cosmologist I have ever known (the late Dennis Sciama) ever changed his mind. I hope that this situation will not continue indefinitely. For cosmology, as in other nefarious pursuits, Sherlock Holmes[4] got it right: *"It is a capital mistake to theorize before you have all the evidence"* (*A Study in Scarlet*, Chapter 3), or *"before one has data, one begins to twist facts to suit theories instead of theories to suit facts"* (*A Scandal in Bohemia*).

References

Ambartsumian, V. A., 1958, Solvay Conference on Structure and Evolution of the Universe (R. Stoops, Brussels) p. 241

Ambartsumian, V. A., 1961, *AJ*, **66**, 536

Ambartsumian, V. A., 1965, Structure and Evolution of Galaxies Proc. 13th Solvay Conference, Univ. of Brussels (New York: Wiley Interscience)

Barris, B. J. *et al.*, 2004, *ApJ*, **602**, 571B

Blanchard, A., Douspis, M., Rowan-Robinson, M., and Sarkar, A., 2003, *A&A*, **412**, 35

Bondi, H. and Gold, T., 1948, *MNRAS*, **108**, 252

Burbidge, G., 1971, *Nature*, **233**, 36

Burbidge, G. and Hoyle, F., 1998, *ApJ*, **509L**, 1

Burbidge, G., Burbidge, M., and Hoyle, F., 1963, *ApJ*, **138**, 873

Dingle, H., 1953, *Observatory*, **73**, 42

Fowler, W. A., Wagoner, R., and Hoyle, F., 1967, *ApJ*, **148**, 3

Freedman, W. *et al.*, 2001, *ApJ*, **553**, 47

Guth, A., 1981, *Phys. Rev. D.*, **23**, 347

Hoyle, F., 1948, *MNRAS*, **108**, 372

Hoyle, F., 1969, *Proc. Roy. Soc.*, **A308**, 1

Hoyle, F. and Narlikar, J. V., 1964, *Proc. Roy. Soc.*, **A282**, 191

Hoyle, F. and Narlikar, J. V., 1966, *Proc. Roy. Soc.*, **A294**, 138

Hoyle, F. and Sandage, A., 1956, *PASP*, **68**, 301

Hoyle, F., Burbidge, G., and Narlikar, J. V., 1993, *ApJ*, **410**, 437

Hoyle, F., Burbidge, G., and Narlikar, J. V., 2000, *A Different Approach to Cosmology* (Cambridge University Press)

Hubble, E., 1929, *Proc. Nat. Science*, **15**, 168

Linde, A., 1982, *Phys. Lett.*, **108B**, 389

Linde, A., 1983, *Phys. Lett.*, **129B**, 177

Mather, J. C. *et al.*, 1990, *ApJ*, **354**, L37

Mather, J. C. *et al.*, 1994, *ApJ*, **420**, 439

McKellar, A., 1941, *Pub. Dom. Astrophys. O.*, **7**, 251

---

[4] I am indebted to Allan Sandage for these quotes.

Narlikar, J. V., Vishwakarma, R. G., and Burbidge, G., 2002, *PASP*, **114**, 1092
Narlikar, J. V. and Burbidge, G., 2004, submitted to *A&A*
Parodi, B. R., Saha, A., Sandage, A. R., and Tammann, G. A., 2000, *ApJ*, **540**, 634
Peebles, P. J. E., 1993, *Principles of Physical Cosmology* (Princeton University Press)
Penzias, A. and Wilson, R. W., 1965, *ApJ*, **142**, 419
Perlmutter, S. *et al.*, 1999, *ApJ*, **517**, 565
Riess, A. *et al.*, 1998, *AJ*, **116**, 1009
Riess, A. *et al.*, 2004, *ApJ*, **607**, 665
Sandage, A., 1963, *ApJ*, **138**, 863
Seife, C., 2004, *Science*, 304, 950
Spergel, D. N. *et al.*, 2003, *ApJ Supp.*, **148**, 175
Steinhardt, P. J. and Turok, N., 2002, *Science*, **296**, 1436
Tammann, G. A., Reindl, B., Thim, F., Saha, A., and Sandage, A. R., 2002, "new era in Cosmology, *ASP Conf. Series*, Vol. 283 (Eds. N. Metcalfe and T. Shanks)
Turner, M., 1993, *Science*, **262**, 861

Discussion

Q : J.-C. PECKER:

I challenge the fact that "expansion" is an observed parameter. The "observed" quantity is the "redshift" z. "Expansion" already implies a theory: Doppler effect is the only redshifting physical process. This is far from obvious, and the door should stay open for tired light mechanisms, and other hypotheses.

A : G. B. :

You are correct. However I maintain that every tired light scheme is incompatible with known atomic physics (basic quantum mechanics).

Comment : M. MOLES :

In your first viewgraph, you listed *the expansion* as one of the observed properties to be accounted for by any cosmological approach. I would not say it is an *observed* property. The "fact" is the Hubble law – the expansion is a theoretical explanation.

In the context of the need for choosing the facts found to be more relevant and the minimum set of hypotheses, even if you agree, it has to be taken into account that right predictions can be obtained from false theories. The case of the Findlay-Freundlich prediction of a BB radiation at $1.6 < T < 6$ from his redshift theory is a good illustration. The key point is to differentiate clearly between well-established "facts" and theoretical explanations and constructions.

Q : H. ARP :

I should propose that the observations show to me that there is a physical continuity between quasars and galaxies with high intrinsic redshifts. It seems to me that there is an evolution between quasars and galaxies, and that galaxy redshifts, and hence expansion of the Universe, is not indicated. For galaxies are created at the same time in a static Universe. One gets the observed Hubble relation quite perfectly.

A : G. B. :
In some instances apparently normal galaxies of stars may have anomalous red-shifts. But the case is not as strong as it is for QSOs and AGN. Most galaxies follow the normal Hubble law. I believe that the difference between them (galaxies and QSOs) is related to the physics of the nuclei. Perhaps QSOs do evolve into normal galaxies after they are ejected from parent galaxies, but we have no theory that will explain this and the observational case is purely circumstantial. The fact is that the vast majority of galaxies follow the Hubble law and this is the most simply understood in terms of expansion.

Comment : J. MORET-BAILLY :
Considering parametric light-matter interactions solves a lot of problems; in particular it explains the periodicities of the redshifts. The "CREIL" (which may be observed with laser, named IRSR) transfers energy between light beams, those that have the largest Planck's temperature being cooled, i.e., redshifted, the others (which are generally thermal beams) being blue-shifted, i.e., amplified. The CREIL requires a low pressure gas, which is not excited permanently. This gas must have a quadrupolar resonance whose frequency is lower than 500 MHz. Atomic hydrogen with a principal quantum number $n = 2$ or 3 works well. It explains all features of the spectra of quasars (gap in the redshifts between the sharp emission lines and the other lines, broad lines if the quasar is radio-quiet, Lyman-α forest, etc.) and shows that the accreting neutron stars have the spectra of the quasars. It explains the blueshift of the radio signals of the Pioneer 10 and 11 probes, and a lot of other observations. The "CREIL" must be considered as a possible solution for many problems.

Q : A. BLANCHARD :
I did not understand your point on the fact that the amount of radiation was chosen by Gamow *et al.* As I understand they picked up the number that explains 25% of helium and *predict* a temperature of a few K. Therefore the discovery of actual black-body radiation is a direct confirmation of the prediction of the model.

A : G. B. :
The point is a very simple one. Gamow found that he had to *choose* a value for the initial photon/baryon ratio to explain the observed helium. If a different ratio were chosen the scheme would not work. Everybody since Gamow has chosen a value close to the one chosen by him. Gamow could not predict the temperature – no

one can. See Turner's article in *Science* and our (Hoyle, Burbidge, Narlikar) book. What Gamow did predict was that the radiation would have black-body form and this is support for the big bang. But the whole thing is based on the *choice* of a ratio, and that is not a theory.

The observed energy density of the CMB 2.7 K is very close to the energy density that is obtained if the helium in the Universe was formed from hydrogen burning in stars. In BB cosmology this is just a coincidence. In our model, this means that hydrogen burning in stars is responsible for the black-body radiation (black-body form is obtained by interaction with dust). Again, read our book.

2

The redshifts of galaxies and QSOs

E. M. Burbidge and G. Burbidge
University of California San Diego, La Jolla, CA, USA

1 Introduction

Redshifts are the lifeblood of cosmology. Until the demonstration that spiral nebulae are external galaxies, and the 1929 discovery by Hubble (Hubble 1929) that there is a correlation of the redshifts with apparent brightness, it was thought that the Universe was static. This is indeed why Einstein put a cosmological constant in his equations. After it appeared that Hubble's result was observational confirmation of the Friedmann–Lemaitre solution to Einstein's equations, it was generally accepted that the Universe is expanding.

Thus from 1930 onwards, there was a strong belief that redshifts could only be due either to Doppler motions away from the observer, as they had been found long ago in galactic stars, or if they were large and systematic they must be shifts due to the expansion of the Universe, and therefore they could be used for cosmological investigations.

Over the years considerable glamour has become attached to the study of large redshifts. This is because they are the only direct tools that give us some measure of what happened in the past, and if we can determine the rate of the expansion H_o we can get some idea of the time scale since the beginning $\sim H_o^{-1}$, i.e., a time scale for the Universe. In this talk we want to give a brief survey of the way that the measurements of redshifts have gone over the last 75 years.

We shall talk first about the major steps taken to measure redshifts of galaxies since Hubble's discovery. In the second part of the paper we shall be concerned with the discovery in 1960 of the quasi-stellar objects or quasars, and show how they have vastly complicated the interpretation of the redshifts. Put succinctly, in our view they have caused a major crisis, because the observations clearly show that the earlier conclusions that redshifts can only be due to Doppler effects or cosmological expansion are not correct.

2 The redshifts and apparent magnitudes of bright galaxies

Following Hubble's work of 1929, in which he used Slipher's redshift measurements, Hubble and Humason, later joined by Mayall, started a program of redshift measurements at Mount Wilson and Lick. The program went on steadily at the Mount Wilson Observatory up to 1950, and started in 1953 at Palomar. By 1936, Humason had obtained redshifts for 146 galaxies, and by 1956, 620 redshifts had been obtained.

A total of 806 galaxies were measured. The vast majority of these are so-called field galaxies with $m \leq 16$. The remainder are galaxies in 18 clusters extending out to $z \approx 0.2$ (Humason, Mayall, and Sandage 1956).

The magnitudes of the field galaxies were measured by Pettit and by Stebbins and Whitford. In this extensive study corrections of the raw data were made both for the redshifts and for the apparent magnitudes. We know from theory that $m_{bol} = 5 \log cz + \text{constant}$.

Humason *et al.* showed that the redshift-apparent magnitude plot for the field galaxies gave a value of the observed slope of 5.028 ± 0.116. The relation was then taken to larger redshifts using the clusters. In this case the curvature must be taken into account through the term q_0.

The final conclusions from this study were as follows.

1. That the slope of the ($\log cz$, m) correlation (for small z) is as close to 5 as the probable errors allow.
2. Unless the effect of evolution of the galaxies can overcome the term involving q_0, the expansion is decelerating. Their attempt to evaluate the galactic evolution gave a value still suggesting that the Universe is decelerating.
3. The expansion appears to be isotropic – this conclusion is simply based on the fact that the 12 clusters in the north and 6 in the south give points on the $\log cz - m$ plot that are indistinguishable.
4. The absolute magnitudes for the brightest galaxies in the clusters are nearly the same as for galaxies in the field.
5. If there is intergalactic absorption, its departure from uniformity must be small, giving only magnitude residuals between zero and 0.3.

Thus some 30 years after the first pioneering discoveries, there appeared to be strong evidence that the Universe is expanding. However the first attempts to measure the curvature of space by Hubble and Humason had failed.

3 The Hubble constant

Hubble and Humason's calibrations of the distance scale were based on the brightest stars in a sample of nearby galaxies. They were thought to be supergiants, and their absolute luminosities were in turn calibrated by using similar stars in M31 and M33, whose distances in turn were determined by the cepheid variables. The zero point of the period-luminosity law for the cepheids went back to the statistical parallax calibration by Shapley and Wilson.

By the 1940s, it had become clear that there were two types of pulsating stars: the classical cepheid variables of Population I and the RR Lyrae stars of Population II, which have two distinct period-luminosity relations. The concept of two stellar populations had been introduced by Baade in 1944. Mineur and Baade and later Blaauw and Morgan showed that this led to a correction of the zero point amounting to $-1.4^m \pm 0.3$ for M31 and M33. This increased the distances of M31 and M33, and hence the absolute luminosities of their brightest stars. A second correction was concerned with the objects that Hubble had originally believed were the brightest stars in nearby galaxies. Sandage began a program with the 200-inch Hale telescope to test this hypothesis. By 1958 Sandage had shown conclusively that Hubble had mistakenly identified H II regions with brightest stars, thus requiring a further correction, again in the sense that the nearby galaxies, in which the brightest star method is used, have distance moduli greater than those estimated by Hubble.

Thus the value $H_o = 558$ km s^{-1} Mpc^{-1} originally obtained by Hubble and Humason was reduced successively to 180 km s^{-1} Mpc^{-1} and then by Sandage to 75 km s^{-1} Mpc^{-1}. By the 1960s there was general acceptance that the value of H_o was much smaller then the original value obtained by Hubble and Humason. In 1962 Sandage, attempting to summarize all of the estimates that had been made by him and others, gave $H_o = 98 \pm 15$ km s^{-1} Mpc^{-1}.

In the 1970s an extensive program to measure H_o more accurately was started by Sandage and Tammann. They calibrated the linear sizes of H II regions as a function of spiral galaxy luminosities, and went on to determine the distances to 39 spiral galaxies of types Sc, Sd, Sm, and Ir using these values. The adopted absolute magnitudes combined with the apparent magnitudes of the Virgo spirals gave a distance to the Virgo cluster of 19.5 ± 0.8 Mpc. The Hubble diagram for first-ranked ellipticals in the Coma cluster scaled back from magnitudes in Coma to magnitudes in Virgo using 5 log cz gave a redshift for the Virgo cluster of 1111 ! 75 km s^{-1}. These values combined to give what Sandage and Tammann called a first hint at the value of $H_o = 57 \pm 6$ km s^{-1} Mpc^{-1}.

Since that time, these authors have used a variety of methods to determine H_o and have concluded that the value has converged and must lie in the range $50 < H_o < 60$ km s^{-1} Mpc^{-1}. One major key to the issue is the distance to the center of mass of the Virgo cluster, which they believe that they have shown by a variety of methods to lie close to 20 Mpc. In Table 2.1 we show how different methods used by them have led to this result. If the true redshift of the Virgo cluster is ~1100 km s^{-1}, a

Table 2.1. *Values of H_o from Sandage (1995)*

Method	H_0 [km s^{-1} Mpc^{-1}]
Virgo Distance	$55 \pm$
ScI Hubble diag	49 ± 15
M101 Diameters	43 ± 11
M31 Diameters	45 ± 12
Tully–Fisher	48 ± 5
Supernovae (B)	52 ± 8
Supernovae (V)	55 ± 8
Unweighted mean	50 ± 2
Weighted mean	53 ± 2

value of $H_0 \sim 55$ km s^{-1} Mpc^{-1} is obtained. Many other methods give a similar result. We show in Table 2.2 a summary of these results of the determinations of H_0 from Sandage.

The two experts, contemporary over the past 20 years with Sandage and Tammann, who were most outspoken in their disagreement with the value of H_0 obtained by Sandage and Tammann, were the late G. de Vaucouleurs and S. van den Bergh. Working actively in the 1970s and 1980s on this problem, de Vaucouleurs (1993) concluded that $90 \leq H_0 \leq 100$ km s^{-1} Mpc^{-1}. In a comparatively recent review, van den Bergh (1992) concluded that $H_0 = 76 \pm 9$ km s^{-1} Mpc^{-1}.

The solution to this problem was chosen to be one of the "key" scientific problems to be solved by use of the Hubble Space Telescope. Both a younger group, involving W. Freedman, R. Kennicutt, J. Mould, and many others, and Sandage, Tammann, Saha, and others have made many observations. But they are still divided. The new group, in the tradition of de Vaucouleurs–van den Bergh, still claim a fairly high value of 71 km s^{-1} Mpc^{-1} for H_0 while Sandage and Tammann still maintain that the value is much smaller. Unfortunately all of the publicity and national attention has been given to the Freedman group alone. This culminated in a press conference in 1999, convened in Washington by NASA, to present the results. Sandage and his colleagues were barred from this meeting. Despite the publicity and the sociological pressure, I believe with Sandage and Tammann that the best standard "candles" appear to be supernovae of Type Ia at maximum light.

Thus a key program has been to find cepheids in galaxies in which those supernovae have been reported and derive their absolute magnitudes M at maximum light. The data have shown that plotting apparent magnitudes of SN Ia against log cz gives a tight relation with a slope of 5, the intrinsic dispersion being $= 0^m.3$. Thus the determination of M (max) should give a good value for H_0, since supernovae

Table 2.2. *QSOs close to bright galaxies ($m_v \le 15.5$)*

Galaxy	m_v	QSO	m_v	z_q	sep$^{(\prime\prime)}$
UGC 0439	14.4	PKS 0038–019	16.86	1.674	72
NGC 470	12.5	(0117 + 0317g)	19.9	1.875	93
NGC 470	12.5	(0117 + 0317g) 68D	18.2	1.533	95
NGC 622	14.0	0133 + 004 (UB 1)	18.5	0.91	71
NGC 622	14.0	0133 + 004 (UB 1)	20.2	1.46	73
IC 1746	14.5	0151 + 048 (PHL 1226)	17.5	0.404	6.4
NGC 1073	11.3	BSO 1	19.8	1.945	104
NGC 1073	11.3	BSO 2	18.9	0.599	117
NGC 1073	11.3	RSO	20.0	1.411	84
NGC 1087	11.5	0243 – 007 (UB 1)	19.1	2.147	170
ZW 0745.1 + 5543	15.3	0745 + 557	17.84	0.174	100
IC 2402	13.5	844 + 319 (4C 31.32)	18.87	1.834	30
NGC 2534	14.0	0809 + 358	18.7	2.40	121
NGC 2693	13.1	0853 + 515 (UB 1)	19.5	2.31	188
UGC 05340	14.8	0950 + 080	17.69	1.45	103
NGC 3067	12.8	0955 + 326 (3C 232)	15.8	0.533	114
NGC 3073	14.1	0958 + 558 (UB 1)	18.8	1.53	144
ZW 1022.0 – 0036	15.5	PKS 1021 – 006	18.2	2.547	122
NGC 3384	10.8	1046 + 129	20.6	0.497	149
NGC 3407	15.0	1049 + 616 (4C 61.20)	16.3	0.422	173
NGC 3561	14.7	1108 + 289	20.0	2.192	66
NGC 3569	14.5	1109 + 357	18.1	0.91	31
NGC 3842	13.3	QSO 1	18.5	0.335	73
NGC 3842	13.3	QSO 2	18.5	0.946	59
NGC 3842	13.3	QSO 3	21.0	2.205	73
NGC 4138	12.1	3CR 268.4	18.1	1.400	174
NGC 4319	13.0	Mk 205	14.5	0.070	43
ZW 1210.9 + 7520	15.4	1219 + 753	18.16	0.645	94
NGC 4380	12.8	1222 + 102 (Wdm 6)	17.6	cont.	88
NGC 4550	12.6	1233 + 125 (Wdm 8)	17.2	0.728	44
NGC 4651	11.8	3CR 275.1	19.0	0.557	210
NGC 5107	13.8	1319 + 38	19.5	0.949	40
ESO 1327 – 2041	13.2	1327 – 206	17.0	1.169	38
ZW 1338 + 0350	14.9	1333 + 0.35	17.98	0.85	41
NGC 5296	15.0	1342 + 440 (BSO 1)	19.3	0.963	55
NGC 5406	13.1	1358 + 932	17.0	3.30	95
NGC 5682	15.1	1432 + 489	19.2	1.940	95
ZW 1640.1 + 3940	15.2	1640 + 396	18.16	0.54	180
NGC 5832	13.3	3CR 309.1	16.8	0.905	372
NGC 5981	13.9	1537 + 595	19.0	2.132	10.7
IC 1417	13.6	2158 – 134	17.8	0.73	76
Anon	15	2237 + 0305	17.3	1.41	≤ 0.3
NGC 7465	13.3	2259 + 157	19.2	1.66	128
NGC 7413	15.2	3CR 455	19.0	0.543	24
NGC 7714 – 15	13.1	233 + 019 (UB 1)	18.0	2.193	120

can be detected in galaxies far beyond the region of the supercluster where perturbations are present, effectively removing most errors in the redshift values.

Galaxies in which supernovae of the so-called "Branch type Ia" have been
identified and have had distances derived now include IC 4182 (SN 1937C),
NGC 5253 (SN 1895B and SN 1972E), NGC 4356 (SN 1981B), NGC 4496
(SN 1960F), NGC 4639 (SN 1990N), and NGC 3627 (SN 1989B). When the
corrections are included they are small and $M_B(max) = -19.47 \pm 0.07$. This
gives $H_o = 56 \pm 4\,km\,s^{-1}\,Mpc^{-1}$ from $M_B(max)$ and $58 \pm 4\,km\,s^{-1}\,Mpc^{-1}$ from
$M_V(max)$.

Since the two values of H_o (the NASA value and the Sandage value) differ by
about 20%, beyond all of the publicity, there is a serious question of to what the
difference is due.

We have concluded that much of the difference should be attributed to the way
in which the Freedman group have used the Tully–Fisher relation. Observational
selection bias (Malmquist bias) has always to be contended with when large numbers of galaxies are observed (cf. Teerikorpi 1997). In that review Teerikorpi shows
a good fit to a value of $H_o = 56\,km\,s^{-1}\,Mpc^{-1}$ based on three basic methods, SN Ia,
Tully–Fisher using averages of several galaxies, and individual cepheid distances.
This is independent confirmation of the results of Sandage and Tammann.

To summarize this section, one of the most fundamental parameters of cosmology
that determines the scale of the Universe had in ~ 50 years been shown to be smaller,
by a factor of approximately 10, than the value originally derived by Hubble and
Humason some 70 years ago.

4 Extension of the Hubble relation to fainter galaxies

By 1972 redshifts of many more clusters had been measured and the Hubble diagram
was extended out to $z \simeq 0.25$. The largest redshift was that for the radio galaxy 3C
295 with $z = 0.461$, which Minkowski had measured in 1960. The relation, which
relied on the brightest elliptical galaxies in clusters, was still strictly linear.

To extend the relation further it was necessary to find fainter clusters. This
was done by photographing random fields with the 200-inch telescope, and by
identifying clusters containing 3C radio sources where it was known that no *bright*
galaxy could be the source. Both methods led to the discovery of fainter galaxies
and hence to larger redshifts.

However, while it is possible in principle to use the Hubble relation to distinguish
between cosmological models, this can only be done if we take into account the fact
that since the fainter galaxies are much further away, we are observing them as they
were when they were much younger than they are now. This requires us to make
corrections for the evolution of galaxies as a function of epoch, using models that,

except at z close to zero, cannot be checked in any detail. Thus this evolutionary correction causes problems if we are trying to determine the deceleration parameter q_0, which measures the departure of the m-log cz relation from linearity.

The general relation is of the form:

$$m_{bol} = 5 \log \frac{c}{q_0^2} \{q_0 z + (q_0 - 1)[(1 + 2q_0 z)^{1/2} - 1]\} + \text{constant}$$

which reduces to

$$m_{bol} = 5 \log cz + \text{constant}$$

for small z.

Bolometric magnitudes are never measured. However, changing to a magnitude in a particular wavelength range only requires a change in the constant in the last equation. This bolometric correction term can be obtained from the spectral energy distribution for galaxies at small z. To convert to objects at high z we must put in a K term, which consists of a term to take account of the fact that the galaxy is no longer being observed in a particular wavelength, but at what for small z would be a shorter wavelength, and also with a bandwidth term 2.5 log $(1 + z)$ to account for the stretching of the spectrum.

Moreover, a galaxy at a high redshift is seen earlier in its life than a nearby galaxy and it has a different absolute luminosity due to the fact that the stars have not evolved as far as those in nearby galaxies. Thus there will be a change in the luminosity, which we call E(z), leading to an apparent magnitude given by

$$m_1 = M_1 - K_1(z) - E_1(z) + 5 \log cq_0^{-2} \{q_0 z + (q_0 - 1)[(1 + 2q_0 z)^{1/2} - 1]\}$$
$$+ \text{constant}.$$

Only for the steady-state cosmological model is E(z) = 0.

Thus, except for the steady-state model, to obtain q_0 from the observed curve, we must be able to determine E(z). This is very difficult. But, for example for $q_0 = 0$, using arguments from stellar evolution, Sandage showed this for ellipticals

$$E(z) = 2.5 \log \left(\frac{L_N}{L_T}\right) \approx \ln \left(\frac{1}{1 + z}\right)$$

where L_N is the luminosity at $z = 0$, and L_T is the luminosity at redshift z. For this cosmological model and a Hubble constant of $H_0 = 50 \, \text{km s}^{-1} \, \text{Mpc}^{-1}$, $E(z, t) = 0.07^m$ per 10^9 years.

A greater uncertainty comes from the possibility that the stellar contents of the galaxies used as standard candles are *not* the same from galaxy to galaxy. Many faint clusters with redshifts out to $z \sim 0.5$ have by now been investigated, and it has become increasingly clear that the complications are increased, so that it is now obvious that the scatter in the data and the evolutionary corrections make it difficult to determine q_0 by this method with any precision. By the 1980s it had been concluded that the method could not be used to determine q_0.

Moreover, Butcher and Oemler (1978) discovered that in at least one faint high-redshift cluster there are many faint blue galaxies. This discovery clearly showed that it could not be assumed that all of the faint clusters simply consisted of galaxies identical to local ($z \simeq 0$) ellipticals seen at earlier epochs. Thus the attempt to determine q_0 by measuring departures from linearity in the Hubble diagram of normal galaxies was abandoned until recently, when it was revised by using Type Ia supernovae in distant galaxies to measure their distances.

The advantage of using Type Ia supernovae in distant galaxies stems from the fact that, as we discussed earlier, they have very similar absolute magnitudes at maximum and thus they are very good "standard candles."

Since the middle 1990s two groups of astronomers have used techniques that enable them to detect large numbers of supernovae of Type Ia in galaxies out to redshifts $z \simeq 0.7$ with the largest redshift detection so far at $z \simeq 1.6$.

All of the Friedmann–Lemaitre models for the Universe with a cosmological constant $\lambda = 0$ predict that the Universe must decelerate as it expands. In the case of the classical steady-state universe or the quasi-steady-state model where matter is continuously being created the Universe will accelerate as it expands. Alternatively, acceleration can be explained if the cosmological constant λ is finite and positive.

The observations by the two groups led to the conclusion that the Universe is accelerating (Perlmutter *et al.* 1999; Riess *et al.* 1998). This was immediately interpreted by these authors as evidence for a positive cosmological constant, and it is now being widely reported that this result has established that the cosmological constant must be non-zero and positive, and there is much talk and belief in the idea that "dark energy" and "quintessence" are dominating the Universe.

No attention is being paid to the perfectly viable alternative, namely that this may indicate that we live in a universe of the quasi-steady-state type (a cyclic universe), as described by Hoyle *et al.* (2000) in which acceleration is predicted. The data are clearly ambiguous even if the theories are not.

Finally in discussing normal galaxies I shall devote a few remarks to the observation of much fainter galaxies thought to lie at much larger redshifts.

5 The fainter galaxies

In order to extend the Hubble diagram to larger redshifts we need to go to much fainter galaxies.

The technique that was used for some time was to find individual galaxies which are powerful radio sources, and obtain optical spectra. At comparatively small redshifts we know that the radio galaxies are predominantly luminous elliptical galaxies and if those could be used as standard candles, the Hubble relation could be extended out to redshifts $\sim$3. However, it turns out, almost without exception, that high z radio galaxies (HZRG) have strong emission line spectra (something that normal giant ellipticals do not have), and thus they may not be simple standard candles.

However some have tried to interpret these systems as mature, slowly evolving galaxies. Up to $z \simeq 2$, the Hubble relation K (infrared luminosity) versus z for the 3C radio galaxies shows a small dispersion. Thus for example, Lilly and Longair argued that if one assumed a large burst of star formation at $z \geq 5$ with declining brightness afterwards, the model was reasonable.

A major difficulty in assuming that the radio galaxies can be used as standard candles is the alignment effect, which was found in the later 1980s. It was shown that the radio, optical, and infrared continua and the emission line structure are all extended and aligned along the major radio axis. This suggests that the event(s) that have given rise to the extended radio sources (the active galactic nuclei) have also given rise to changes in the luminosity of the galaxies, thus making it less likely that they can be used as standard candles.

In the last decade a great deal of work has gone on to try to understand the population of faint galaxies and to measure their redshifts, and I shall be very brief in summarizing this work. The major drives have been (a) because the newer telescopes and detectors have enabled us to go much fainter than ever before, and (b) because it is generally believed that in looking back in redshift and sampling galaxies of an earlier epoch, we are getting closer to the time when all galaxies were young.

Unfortunately this latter argument depends very strongly on the belief in the idea that all galaxies formed soon after a big bang, which may well not be true. That this is the case can be seen from the earliest observational attempts to find primeval galaxies by Partridge and by Davis and Wilkinson, who were following the initial prediction of Partridge and Peebles.

Early attempts were made by Djorgovski to detect redshifted Lyα. While some Lyα emitters were seen close to high redshift QSOs, as we shall discuss in the second part of this paper, it may well be that the QSOs are not trustworthy when it comes to using their redshifts as cosmological probes.

The major breakthrough in detecting high redshift galaxies came with the realization within the UV spectra of star-forming galaxies that there will be a Lyman continuum discontinuity at 912A. This is called the Lyman break. The use of this, pioneered by Steidel and Steidel with others (Steidel *et al.* 1996) had led to the measurement of many hundreds of galaxy candidates with redshifts in the range $2.2 < z < 3.6$ using ground-based telescopes and HST. An extensive review has been given recently by Giavalisco (2003). In order to confirm that such galaxies are star-forming but otherwise normal galaxies, and not AGN or QSOs, individual spectra are required. This requires the largest telescopes (8–10 meter class). By now large spectroscopic samples are available over the redshift range $2 < z < 3.5$. A limited number of redshifts have been found of galaxies in the Hubble Deep Field. Here there are some Lyman break galaxies with $z > 4$. There have also been some studies of the faint red galaxies. When such galaxies were first discovered they were thought to have very high redshifts ($z = 20$), but later they were thought to have $2 < z < 3$, and by now it has been concluded that the redshifts are smaller, in the range $z = 1$–2. A recent review has been written by McCarthy (2004).

All of the studies of the faint galaxies have been concerned with their colors, luminosities, and chemical compositions, and whether or not evolution can be detected and how they fit into the popular schemes of galaxy formation. Not surprisingly, they have not really told us very much about the Hubble diagrams for the faintest detected systems.

This is all I want to say about the redshifts of "normal" galaxies. However before I turn to the redshifts of QSOs I want to mention two aspects of galaxy redshifts that have not had much attention paid to them, but may ultimately be important.

6 Anomalous redshifts of galaxies

a. **Small periodic effects in Δz.**

Tifft (1976) first pointed out that if the differences between the individual redshifts of galaxies in a rich cluster, the Coma cluster, are plotted it appears that the differences tend to show a periodicity with the period $c\Delta z \simeq 72$ km s^{-1}. While this result was confirmed by Weedman, and found in other clusters, the effect has been largely disregarded. However the most precise study of this by Guthrie and Napier (1996) using very local spiral galaxies with very accurate redshifts, measured using the 21 cm line, has confirmed the reality of the effect, though they have concluded that the periodic term $c\Delta z = 37.6$ km s^{-1}.

It remains unexplained.

b. **Anomalous redshifts of pairs or multiple groups of galaxies.**

For more than thirty years we have known that there are many tight groups of galaxies with at least one redshift that is far greater (or smaller) than the others (Burbidge and Burbidge 1959; Burbidge and Sargent 1971; Arp 1987; Arp and Sulentic 1985; Hickson 1997 and many other references).

Of course most astronomers have ignored these results, and if told about them have refused to believe the statistical evidence. In other words, they want to attribute all such configurations to accident. So once again the data have been ignored, but I believe that they are real. Dr. Napier will talk about the first of these topics.

7 QSOs: Introduction

These objects were first identified in 1960 as radio sources and after only a few had had their redshifts measured it was becoming clear that there was little, if any, correlation of z with apparent magnitude (see Fig. 2.1, taken from Hewitt and Burbidge 1993). This, together with the fact that they were soon found to vary in time both in radio flux and in optical flux, showed that they were very different from normal galaxies of stars. Of course, their spectra show no indication at all that stars are present, though their chemical composition based on emission line studies suggested that the chemical composition is fairly normal (Hamann and Ferland 1999).

The variability meant that they are very small, probably no bigger than the solar system, and this led to difficulty with the physics if the redshifts were measures of their distances. But despite these problems most astronomers assumed that the very large redshifts (3C 9 with $z = 2.012$ had been found in the first 20) meant that they could be used for cosmological investigations.

However, the strongest evidence that, at least for some of them, this could not be the case, came through studies of their proximity and apparent associations with comparatively nearby bright galaxies. Figure 2.2 shows an interesting case of such a connection. Unfortunately, only a few astronomers chose to investigate this aspect of the problem but, after more than thirty years, we believe that the case for the existence of local QSOs whose redshifts in large part are not due to the expansion of the Universe has been established.

In the next sections we will outline this evidence.

8 The data

It was Arp (1966a, b, 1987) who first showed that many of the powerful radio sources that had been found in the 3C and Parkes radio source surveys appear to lie in pairs across peculiar interacting galaxies in his catalogue. Some of these sources had been identified with high redshift QSOs.

Soon after this a proper statistical study was made of the positions of the complete sample of 3C QSOs (50 in all) and the galaxies in the Shapley–Ames catalogue. In this study it was found that five of the QSOs lie within 6 arcminutes of bright galaxies, and Monte-Carlo tests showed that there was a very strong probability that these associations are real (Burbidge *et al.* 1971). See further discussion of these data by Kippenhahn and de Vriess (1974). Arp investigated the environments of many bright

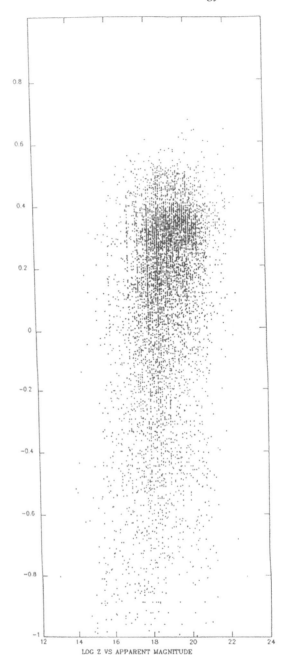

LOG Z VS APPARENT MAGNITUDE

Figure 2.1 Redshift – apparent magnitude plot for 7315 QSOs taken from the catalogue of Hewitt and Burbidge (1993).

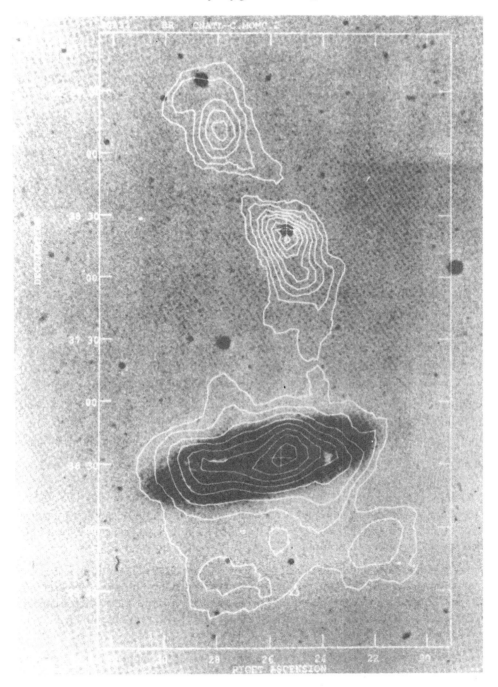

Figure 2.2 NGC 3067, showing 21-cm radio contours and the QSO 3C 232 ($z =$ 0.533) at the center of the 21-cm contour immediately north of (above) the Galaxy (from Hoyle *et al*. 2000).

galaxies and showed that there were many cases of close proximity with QSOs with very low probability that these are accidental configurations (Arp 1987).

In Table 2.2, we showed a large number of such cases found by Arp and others as listed by Hoyle *et al.* (2000). Even stronger evidence for physical connection between a high redshift QSO and a galaxy is present in a few cases – e.g., NGC 4319 and MK 205, and the connections between the galaxies in NGC 7603 (Arp *et al.* 2003). Again, these were found by Arp, and in recent years Lopez-Corredoira and Gutierrez have found that in the case of NGC 7603 there are indeed four objects all with very different redshifts connected by luminous bridges.

In the last ten years or so the few people working in this field have used the X-ray surveys to identify QSOs, which often are clustered about active galaxies (AGN).

9 QSOs associated with AGN galaxies

A program was begun with observations with the ROSAT satellite by Pietsch *et al.* (1994), and examination optically of the ROSAT PSPC fields by Arp (1997) and Radecke (1997) showed excess numbers of point X-ray sources within the 50 arc min fields centered on the AGN galaxies. This is very clearly shown in figures published by Radecke (1997); he determined the probability that the effect is real as $\sim 7.6\sigma$. Arp and Radecke, working with the ROSAT data, found a number of coincidences between ROSAT sources and faint blue stellar objects (BSO), and published coordinates and finding charts for several of them. In the last few years a number of examples have been studied. I shall describe some of them in what follows.

NGC 4258

Two BSOs were found to lie roughly symmetrically on each side of this nearby active galaxy ($D \simeq 4$ Mpc) (Pietsch *et al.* 1994). He applied for observing time at the optical ESO telescope at Calar Alta, but did not obtain any. This is shown in Figure 2.3.

During a visit to Garching, Pietsch and Arp showed E. M. Burbidge the data, and asked whether I could observe the BSOs, which were candidate quasars. She had some observing time at Lick Observatory requested, so she agreed, and she used the Lick 3 m telescope to obtain spectra of the BSOs. They were both found to be QSOs with redshifts 0.398 and 0.653 respectively (Burbidge 1995).

NGC 4258 is a very interesting galaxy in itself; it has complex motions of the gas in its central region, H_2O megamaser measurements have shown central rotation around a massive black hole, and multiple-valued velocities in the gas at the center of NGC 4258 have been found. It clearly merits further detailed study.

Arp and colleagues followed the study of the QSOs apparently associated with NGC 4258, with observations of optical identifications of BSOs at ROSAT X-ray positions around a

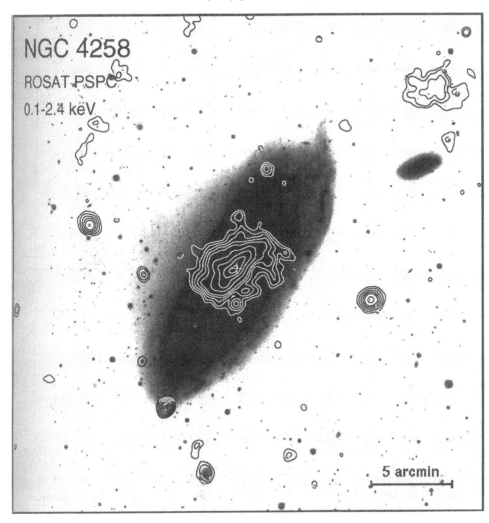

Figure 2.3 Field of NGC 4258 (from Pietsch *et al.* 1994) showing the two QSOs at the positions of ROSAT point sources.

number of AGN galaxies, of which we give details for the galaxies NGC 2639, NGC 3516, NGC 1068, NGC 3628, M82, Arp 220, and NGC 7319.

NGC 2639
The next Arp/Radecke field with candidate QSOs aligned almost on the minor axis is NGC 2639, where there are several known QSOs in the field (Arp 1997). Lick observations of the NE ROSAT candidate QSO (Burbidge 1997) showed it indeed to be a QSO: thus there are two ROSAT QSOs aligned almost on the minor axis and at almost equal distances from the nucleus of this AGN galaxy, with almost equal redshifts $z = 0.3048, 0.3232$. There is also a line of seven less strong X-ray sources conspicuously extended NE from the nucleus

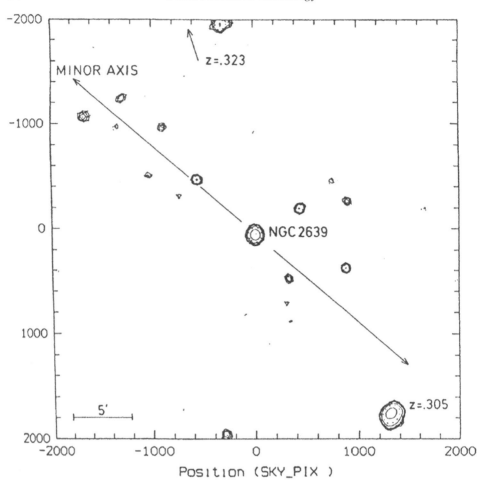

Figure 2.4 Field of NGC 2639, showing line of QSOs along its minor axis, from Burbidge *et al.*, in press.

of NGC 2639, along its minor axis (see Fig. 2.4). Our recent observations at the Keck Observatory have shown them all to be QSOs, with $z = 1.304, 0.352, 2.630$, and 0.337.

NGC 1068

This Seyfert galaxy has also been shown to be at the center of a group of 11 QSOs brighter than $V = 19$ mag., with redshifts $z = 0.261$ to 2.108, lying within a radius of $50'$ of the galaxy. These data are shown in Burbidge (1999).

NGC 3516

Chu *et al.* (1998) noted a very interesting configuration of five X-ray emitting BSOs distributed in a line along the minor axis of the Seyfert galaxy NGC 3516, which is a strong X-ray emitting galaxy in the ROSAT list. Spectra they obtained with the 2.16 m telescope at Xinlong Station, Beijing Astronomical Observatory, showed the five objects to be QSOs with redshifts 0.33, 0.69, 0.93, 1.40, and 2.10. This configuration, strongly suggestive of the ejection of these QSOs from the nuclear region of NGC 3516, is one of the most interesting configurations discovered in the study of QSOs in the fields of AGN galaxies.

NGC 3628

This is a starburst/low-level AGN galaxy in the Leo Triplet, noted for its extensive outgassed plumes of neutral hydrogen, which is undergoing major internal dynamic activity. The field around NGC 3628 is particularly interesting, because of the large number of QSOs within a few tens of arc minutes of the nucleus of the galaxy, and a tail of X-ray emission extending 4 arc min south, and linking QSOs with $z = 0.995, 2.15, 1.75$ to the nuclear region of NGC 3628; Fig. 2.5 shows this. Nine QSOs ($z = 1.46, 2.06, 1.75, 2.15, 0.995, 0.981, 0.408, 2.43, 1.94$) lie within a circle of radius 17 arc min of the center of NGC 3628. Within $\sim 15'$ of its center there are seven known QSOs and one probable QSO, and the two QSOs a few arc min south of the nucleus are connected to the nucleus, as shown by a broad band ROSAT map smoothed and contoured from PSPC photon event files. An R-band exposure with VLT-FORS2 shows this chain of objects (ref. Arp *et al.* 2002; *ESO Messenger* March 2002).

M82

This nearly edge-on galaxy was shown in early studies to have outflowing gas perpendicular, north and south, to the major axis of this starburst galaxy. This outflow is now known to be related to the tremendous star-forming activity in its central regions. Studies from 1980 on have shown: (a) nine QSOs lie in a cone extending about $10'$ SE of M82; (b) a filament NE was detected in Hα emission by Devine and Bally (1999); (c) 17 unresolved X-ray sources were mapped by Dahlem *et al.* (1998) outside, but close to, M82, of which they suggested "that at least some of the unresolved sources are associated with M82." We were able to measure with the Keck 10-m Telescope stellar objects, which Arp had identified at six of the Dahlem *et al.* positions, and found all to be fairly low-redshift QSOs (Burbidge *et al.* 2003), where we also list the other QSOs, already known, close to M82.

Arp 220

Arp 220 is a nearby ultraluminous infrared galaxy that is a strong X-ray source. Four compact ROSAT X-ray sources close to Arp 220 identified with blue stellar objects were observed at Lick Observatory and Beijing Astronomical Observatory. All are QSOs. The two closest to the center have almost equal redshifts 1.249, 1.258, and the outer two have $z = 0.232, 0.463$ (Arp *et al.* 2001). This field merits further study; Arp 220 is a complex object.

NGC 7319

So far, we have presented data on QSOs close to, but outside, the galaxies with which they are associated. Our most recent data have been on an optical object at the position of one of the ultraluminous X-ray sources (ULX), which are being discovered by Chandra and XMM-Newton lying *within* spiral galaxies. Since they emit at levels above $10^{38.5}$ erg s^{-1} these cannot be normal X-ray binaries. Burbidge *et al.* (2003) suggested that they might be QSOs, with a wide range of redshifts, in the process of being ejected from the galaxies.

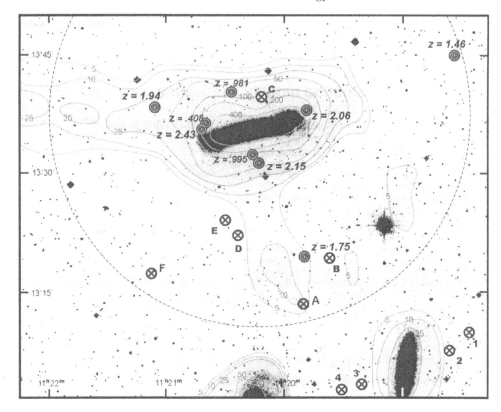

Figure 2.5 NGC 3628, showing isophotal contours of neutral hydrogen and loca-
tions of the numerous known QSOs and candidates as yet not observed, from Arp
et al. (2002).

One such object has been found only eight arc seconds from the nucleus of NGC 7319, a
Seyfert galaxy in Stephan's Quintet. With the Keck 10-meter telescope we have shown that
it is a QSO with $z = 2.11$ (V = 21.79) (Galianni *et al.* 2005).

10 Conclusion

This is a brief survey of the work on determination of the Hubble constant from
the redshifts of galaxies, and the work now ongoing on the measurement of the
large redshifts of faint, very distant galaxies. Then we discuss the problem of the
redshifts of the QSOs, with examples of some galaxies with active nuclei where
QSOs with very different redshifts appear to be associated with these galaxies. When
the number of such QSOs in the small fields around such galaxies are compared with
statistical work on numbers of QSOs in the general field – such as work with the 2dF
instrument on the Anglo-Australian Telescope – the concentration of QSOs around

the AGN galaxies is many times the distribution in the general field, indicating that the associations are real. The problem facing us – observers and theoreticians – is to understand the physics of QSO redshifts. If the QSOs are formed in or very near the nuclei of the galaxies, new physics is required. The spectra of these QSOs show the presence of elements heavier than hydrogen, He, C, N, O, Fe, whose formation in stellar interiors and in supernova explosion is the subject of many studies. Are the QSOs portions of the nuclei of AGN galaxies, which some as-yet not understood energetic process causes to be ejected with rather high velocities? One of the first such associations of X-ray-emitting QSOs, 9′ from the nucleus of NGC 4258, is a radio source, so the very high resolution radio measurements may reveal in several years whether it has a proper motion away from the galaxy.

In conclusion, I want to thank Drs. Pecker and Narlikar for organizing this wonderful conference, and Shirley Roy for all her help with the manuscript and the figures.

References

Arp, H. C., 1966a, *Atlas of Peculiar Galaxies* (Calif. Inst. of Technology, Pasadena)

Arp, H. C., 1966b, *ApJ Supp.*, **14**, 1

Arp, H. C., 1987, *Quasars, Redshifts and Controversies* (Berkley, & Interstellar Media), **157**, 8

Arp, H. C., 1997, *A&A*, **319**, 33

Arp, H. C. and Sulentic, J. W., 1985, *APJ*, **291**, 88

Arp, H. C., Gutierrez, C. M., and Lopez-Corredoira, M., 2003, *A&A*, **418**, 877

Arp, H. C. *et al.*, 2001, *ApJ*, **553**, 611

Arp, H. C. *et al.*, 2002, *A&A*, **391**, 833

Burbidge, E. M., 1995, *A&A*, **298**, L1

Burbidge, E. M., 1997, *ApJL*, **484**, L99

Burbidge, E. M., 1999, *ApJL*, **511**, L9

Burbidge, E. M. *et al.*, 2003, *ApJ*, **591**, 690

Burbidge, E. M. and Burbidge, G., 1959, *ApJ*, **130**, 23

Burbidge, E. M. and Sargent, W. L. W., 1971, *Pontif. Acad. Sci. Scr. Varia*, No. **35**, 351

Burbidge, G. R., Burbidge, E. M., Solomon, P. M., and Strittmatter, P. A., 1971, *ApJ*, **170**, 233

Butcher, H. and Oemler, A., 1978, *ApJ*, **219**, 18

Chu, Y. *et al.*, 1998, *ApJ*, **500**, 596

Dahlem, M., Weaver, K. A., and Heckman, T. M., 1998, *ApJ Supp.*, **118**, 401

de Vaucouleurs, G., 1993, in *Physical Cosmology* (Ed. D. Schramm, Washington, D.C.)

Devine, D. and Bally, J., 1999, *ApJ*, **510**, 197

Galianni, P., Burbidge, E. M., Arp, H. C., Junkkarinen, V., Burbidge, G., and Zibetti, S., 2005, *ApJ*, **620**, 88

Giavalisco, M., 2003, *ARAA*, **40**, 579

Guthrie, B. N. G. and Napier, W. M., 1996, *A&A*, **310**, 353

Hamann, F. and Ferland, G., 1999, *ARAA*, **37**, 478

Hewitt, A. and Burbidge, G., 1993, *ApJ Supp.*, **87**, 451

Hickson, P., 1997, *ARAA*, **35**, 357

Hoyle, F., Burbidge, G., and Narlikar, J. V., 2000, *A Different Approach to Cosmology & Cosmogony* (Cambridge University Press)
Hubble, E., 1929, *Proc. Nat. Acad. Sci.*, **15**, 168
Humason, M. L., Mayall, N. U., and Sandage, A. R., 1956, *AJ*, **61**, 97
McCarthy, P., 2004, *ARAA*, **41**, (in press)
Perlmutter, S., *et al.*, 1999, *ApJ*, **517**, 565
Pietsch, W., Vogler, A., Kahabka, P., Jain, A., and Klein, V., 1994, *A&A*, **284**, 386
Radecke, H. D., 1997, *A&A*, **319**, 18
Riess, A., *et al.*, 1998, *AJ*, **116**, 1009
Sandage, A., 1995, *The Deep Universe* (Saas-Fee Advanced Course 23), (Eds. A. R. Sandage, R. G. Kron, and M. S. Longair, Berlin, Heidelberg, Springer), p. 77
Steidel, C. C., Giavalisco, M., Dickinson, M., and Adelberger, K. L., 1996, *AJ*, **112**, 352
Teerikorpi, P., 1997, *ARAA*, **35**, 101
Tifft, W. G., 1976, *ApJ*, **206**, 38
van den Bergh, S., 1992, *PASP*, **104**, 861

Discussion

Q : W. NAPIER:

I wonder whether the apparent extension of the X-ray contours from NGC 3628 to the QSOs at $z = 0.995$ and 2.15 might be an artefact, caused by the "bleeding" of separate extended X-ray fields? The same question might be asked of the radio map of 3C 343.1, showing an apparently common bridge between it and the nearby QSO.

A : M. B. :

I do not think that point X-ray sources can "bleed" in this way. I was impressed by the large-telescope image of the S part of NGC 3628, which we showed in our paper. Also, the large number of QSOs around NGC 3628 is unusual, given the standard counts of QSOs per square degree. In the case of 3C 343.1, we are looking at radio contours, not X-ray contours, and I think it unlikely that radio contours can be considered to "bleed" in this way.

3

Accretion disks in quasars?

Jack W. Sulentic

University of Alabama, Tuscaloosa, 35487 USA

Abstract

This polemic considers the reality and implications of broad double-peaked Balmer, and super-broad asymmetric FeK alpha, emission lines in quasars. Current evidence suggest that both are rare. The lack of physical consistency and/or correlation in a disk model parameter space suggests little support for the claims that these lines arise from an accretion disk.

1 Introduction

Everyone knows that the central engine of a quasar involves accretion onto a super-massive black hole (SMBH). What else can it be? Especially if the Doppler interpretation of quasar redshifts is accepted and is correct. What else could the redshift be? This ideological approach to science is both good and bad. Without a paradigm research in this area would lack any focus or direction. Such anarchy is clearly out of favor. The danger, however, is that a paradigm can be confused as fact and efforts to explore, or even hypothesize, alternatives is discouraged and even suppressed. This tendency can be re-enforced in more ideologically oriented cultures because scientists, after all, are not immune to the weaknesses of the societies in which they work.

There is a tendency, when a paradigm becomes too strong, for observations to be treated with a measure of skepticism and even contempt. This is true *unless* they support the prevailing beliefs. Part of the disbelief can stem from genuine skepticism given the difficulty of obtaining good data. 'Good' is obviously an ill-defined term but, in this context, it involves a clear understanding of what a given set of data can, and cannot, tell you. In other words the ability of that data to constrain models. Quasar spectroscopy is a good example of this tendency. It has been out of fashion for more than a decade because of both: 1) the difficulty of obtaining spectra which reasonable S/N and resolution, and 2) the failure of models to reproduce what is observed. There may well be an added fear component injected by Arp and collaborators: the fear that something really challenging to the paradigm might be discovered. While discovery is exciting to a scientist, it can be a source of fear to a careerist.

Following the above lines, any single piece of data that supports the paradigm will often be accepted without question and lionized in the scientific press. Excellent recent examples involve: 1) double-peaked Balmer emission lines in the optical spectra of some quasars and 2) broad 6.4 keV FeKα emission lines in the hot, young field of quasar X-ray spectroscopy. We use the word quasar here in a generic fashion, which means all classes of extragalactic objects that show broad (full width at half maximum FWHM 1000 $\gtrsim$ km s^{-1}) optical emission lines. This includes the nuclei of Seyfert and radio galaxies as well as radio-quiet and radio-loud point source quasars. The galaxies tend to show redshifts $z < 0.1$ while the latter show a redshift distribution that peaks near $z = 2.0-2.5$. They are often united under the umbrella of "active galactic nuclei" (or "AGN"), which re-enforces the paradigm that they are driven by the same physics, differing only in their redshift-implied distances and, hence, luminosities.

AGN with both double-peaked Balmer and broad FeKα emission lines have been widely touted as the direct signature of a line-emitting accretion disk. Why were such observations so widely hyped? For example, both were editorialized in *Nature*. Perhaps that should be a warning sign? No attempts to refute, or even moderate, this view appeared there. The reason for this hype stems from the implication – direct detection of a line-emitting accretion disk (AD) is as close as we will ever come to proving the existence of SMBH. AD and SMBH may well exist but do these observations constitute the proof they are claimed to represent?

Consideration of these two independent lines of spectroscopic evidence is not directly related to the subject of this conference – as Geoff Burbidge pointed out. But perhaps Geoff is affected too much by the anti-paradigm effect – he does not require observational evidence to "know" that the standard paradigm is wrong. Since the organizers asked me to speak on "accretion disks," and I doubt that they want to hear the latest models emerging from the burgeoning industry of accretion disk theory, it seems worthwhile to look at these data with a hard eye. After all, if the existence of AD and SMBH can be proven then maybe there was reason to organize this conference. The data should be allowed to speak and surely they should cut the AD models both ways. This is at least a reasonable approach if there is still a need for empiricism in science. In the next two sections let us consider, in the hard light of day, the observations that were obtained (at least half of them) in the dark of the night.

2 Double-peaked Balmer emission lines

Considerable excitement was generated when reasonably high S/N spectra of an AGN, appropriately named Arp 102b, revealed very broad and double-peaked Balmer emission lines. This discovery was quickly followed by demonstrations that the lines could be reasonably well fit by the sort of profiles expected from a relativistic Keplerian AD (Chen and Halpern 1989; Chen, Halpern, and Filippenko 1989). Other bright and famous sources like Pictor A (Halpern and Eracleous 1994), 3C390.3 (Perez *et al.* 1988), and 3C332 (Halpern 1990) also show such double-peaked line profiles. At face value the evidence for AD line emission looks good but let's see how far that goes (see also Sulentic *et al.* 1999) and, especially, let us contextualize the data and the model. After all, quasars show such a spectral diversity that almost anything can be found. The paradigm, however, holds that all quasars are powered by an AD (fresh supply + SMBH (central engine).

- While the above objects are intriguing, the fact is that double-peaked line profiles are very rare. The above discoveries were followed up by a campaign to find more. Considerable time on large telescopes was granted (as inferred from the published papers) to carry out a spectroscopic search. Since all, or most, of the known double-peak sources were radio-loud, the search concentrated on radio-loud AGN (about 10% of AGN are radio-loud) with broad and/or complex Balmer line profiles (Eracleous & Halpern 1994; 2003). Perhaps ∼20% of the sources showed some kind of twin shoulders but only ∼60% of those could be well fit with profiles derived from simple relativistic disk models. This leaves us with less than 1% of all AGN that show such spectra, which leaves us with three possible interpretations of the double-peaked sources. They are either: a) miraculous, b) pathological, or c) observed at a special orientation to our line of sight. Our own work involving spectra of 200+ bright, low redshift AGN (Marziani *et al.* 2003) suggests that these sources are outliers, i.e. rare and/or hathological, when compared with the distribution of Balmer line FWHM for a large sample of quasars. If they are rare because we view them at a preferred orientation, then they should statistically favor sources with AD oriented near face-on rather than edge-on. But sources wtih double-peaked lines show the BROADEST profiles observed implying intermediate or edge-on viewing angles where the MAJORITY of sources should be observed. The Broad lines in sources with AD oriented near edge-on are throught to be observed by an optically thick torus (called type 2 AGN). The preferred orientation interpretation would instead favor models involving binary SMBH or bicone outflows (e.g., Zheng *et al.* 1990; Sulentic *et al.* 1995). The former has failed basic observational tests (Eracleous *et al.* 1997) while the latter has problems and, even if correct, offers no support for the AD paradigm.
- Radio-loud sources, interpreted as the AGN with the most directly visible AD, show the weakest optical FeII line emission. FeII emission can be argued to be one of the defining characteristics of broad line AGN. Ironically the ubiquity and strength of FeII emission in AGN is perhaps the strongest argument in favor of the AD paradigm. An AD provides the necessary physical conditions (e.g., high electron density and column density; Dumont

& Collin-Souffrin 1990) to account for the FeII emission. If FeII emission arises in an AD then double-peaked lines probably have nothing to do with an AD.

• Several of the well known double-peaked sources have been monitored for variability and the observations reveal that the two emission lines peaks vary out of phase (Arp 102b – Miller & Peterson 1990; 3C390.3, Zheng *et al.* 1991). Others show transient double-peaked line structure (NGC 1097 – Storchi-Bergmann *et al.* 1997; Pictor A – Sulentic *et al.* 1995). Out-of-phase variability requires more exotic (i.e., more free parameters) disk models and transience is inconsistent with a stable AD.

• The initial hype involving Arp 102b motivated us to try a more general confrontation between disk models and line profile parameters (Sulentic *et al.* 1990). For example, simple relativistic Keplerian disk models predict double-peaked profiles with: a) a stronger blue peak (from Doppler boosting) and b) a redshifted base (from gravitational redshifting at the innermost radii). We explored a wide range of AD illumination models and also took source orientation into account. Model predictions showed no general agreement with observation. Given the growth of the AD model industry, it is safe to say that a model can be found to fit almost any conceivable profile, but that is model fitting and not science. One gets the impression than this is ok after a paradigm has become "fact". It is obvious that real support for the disk paradigm requires some model fitting, a sample of sources, and finding some concentration or correlation of the individual fits in the n (= 4 or 5) dimensional AD model parameter space. The fits to the radio-loud sample mentioned above (Eracleous & Halpern 1994; 2003) show no evidence for such a correlation. Individual source fits scatter widely in the parameter space without any hint of correlation with observed source properties. Where is the Nysics

So where are we 15 years after the double-peak hype began? The rarity of double-peaked sources remains a thorny issue. This, and rigid adherence to the paradigm, likely motivated a search for double-peaked sources among the plethora of quasars (more than 3000 searched) in the SDSS archive. This search yielded 116 sources ($\sim$3–4%) with "double-peaked" Balmer lines among the quasars (Strateva *et al.* 2003). Most of the candidate SDSS double-peaked quasars are radio-quiet in contrast to the earlier survey that found most double-peakers to be radio-loud. That is potentially good for the AD paradigm because radio-quiet quasars represent the majority and they should host AD+SMBH too. But are they double-peaked sources? A glance at the candidate atlas reveals that most are not double peaked. Very few look like the sources in Eracleous & Halpern (1994; 2003). Many can be described as lumpy profiles. In many cases one of the lumps is red- or blue-shifted relative to the inferred quasar rest frame while the other peak is unshifted. In a number of cases a central single-peaked component is subtracted from the broad profile to yield a more double-peaked appearance! It is true that most quasars show single-peaked and relatively unshifted line emission. It is true that this more ubiquitous line component might "get in the way" of seeing the double peaks. But such data modification would be loudly denounced if it were done in support of anything outside the paradigm. Line profiles in most or all of the 116 sources were fit with disk models. As we noted before, one can find a disk model to fit almost any line profile and this is the proof of it. Needless to say, the parameter fits fill the entire n-dimensional disk parameter space. There is no physics here. Any connection between double-peaked line profiles and AD physics is weak or nonexistant.

3 The FeKα emission lines

X-ray spectroscopy of quasars is only about 10 years old. A K-shell iron line at 6.4 keV is the strongest emission line feature observed in the X-ray spectra of quasars. The ASCA mission permitted the first line profile measures of this FeKα line in many AGN. Considerable excitement greeted the relatively high S/N observations of the FeKα line in the Seyfert galaxy MCG-6-30-15 (Tanaka *et al.* 1995). The excitement was greater than for Arp 102b (in 2004: 340+ citations vs. 120–140) because, while not double-peaked, the FeK profile was extremely broad and highly asymmetric. It showed a sharp narrow peak on the high energy (blue) side and a red extension implying velocities up to 100 000 km s^{-1}. It was immediately fit with a disk model that suggested we were seeing line emission from the innermost radii of an AD. This discovery motivated a flurry of AGN spectroscopic observations of other AGN with ASCA (Nandra *et al.* 1997a, b; Reynolds 1997). An enormous number of trees were cut down to fuel the multitude of papers published during this paradigm driven hysteria. The conclusion in the end was that the FeKα line was broad and complex in all or most low redshift sources. And, of course, it was assumed to arise from the AD in all sources. Let us examine these data in the cold light of day (see also Sulentic *et al.* 1998a, b).

Echoing the results for the double-peaked Balmer lines, the Fek profile fits showed no convergence in a disk parameter space (or Kerr vs. Schwarzschild metric for that matter) with one possible exception – most of the data required line emission from the innermost radii of the AD. The majority of FeK profiles showed a blue peak that always agreed closely with the rest frame of the quasar (i.e., 6.4 keV). It was easy to show that a modest range of source orientations to our line of sight (e.g., 15°) would produce a considerable scatter in the peak energy of this narrow component interpreted as the Doppler boosted wing of a relativistic disk profile. The constancy suggested to us that the broad redshifted and narrow unshifted components of the line were independent (in MCG-6-30-15 they vary out of phase; Iwasawa *et al.* 1996). In this case the unshifted narrow component offers little hope as a proof of the AD paradigm. We argued that decomposition of the profile left a broad redshifted component that was as well fitted by a symmetric Gaussian as by a more complex shape. Naturally the former symmetric fit is incompatible with disk emission from a relativistic regime. Our citation rate did not fare as well as the earlier ones but, when one considers the fate of bearers of bad news in earlier times, this is a small price to pay for playing the devil's advocate.

So where are we almost ten years after the FeK hype began? It is a very different world because we have CHANDRA and NEWTON, which provide spectra of much higher S/N and resolution. The biggest change is that many of the sources thought to show the broad complex FeK profile are not confirmed with better data (about with, vs 55 without, at. Low S/N and resultant poor definition of the underlying continuum plagued the ASCA spectra. Now we often see sources with the relatively narrow spectral signature of cold, warm, and hot FeK at 6.4, 6.7, and 6.9 keV respectively. MCG-6-30-15 remains the source with the strongest broad FeK line. Some adopt our (and a few others) two-component interpretation of the line and move the redshifted component inside the last stable orbit (Wilms *et al.* 2001), others (Vaughn and Fabian 2004) continue to insist that the line is an AD signature, and still others abandon the AD paradigm in whole or part (e.g., Misra 2001; Inoue and Matsumoto 2003). Many papers (e.g., Page *et al.* 2003) now openly admit that the FeK line no longer provides strong direct evidence for the relativistic AD paradigm. In the unlikely possibility that MCG-6-30-15 does provide such evidence, it is of minor importance because it can no longer be generalized to the majority of AGN.

4 Conclusions

The claims that double-peaked optical and 6.4 keV X-ray emission lines arise from an AD in the majority of AGN do not survive close scrutiny. No physics surrounds these claims – only ideology and unrestricted AD model fitting. This does not prove that disks do not exist, only that direct evidence for them does not lie with these data. Why is this worth a presentation at an unconventional meeting? Why is the message presented so strongly? In part it is intended as a counterpoint to the strident claims that have been made in the past 15 years in support of the AD paradigm. There is also the hope that students will become more sensitive to the role that paradigm plays in their interpretation of data. Perhaps it will encourage them to be more skeptical and to consider alternatives, at least mainstream ones, when evaluating data in the context of the standard model. This is best done after they have found a permanent job. Finally the message to the few people working on real alternatives is that they should not be intimidated by the edifice of data that is presented (even at this meeting) in support of the standard big-bang model. The evidence is composed of many pieces and many of them, when examined carefully, will have all of the weaknesses of the ones discussed here.

References

Chen, K. & Halpern, J., 1989, *ApJ*, **344**, 115
Chen, K., Halpern, J., & Filippenko, A., 1989, *ApJ*, **339**, 742
Dumont, A. & Collin-Souffrin, S., 1990, *A&A*, **229**, 313

Eracleous, M. *et al.*, 1997, *ApJ*, **490**, 216
Eracleous, M. & Halpern, J., 1994, *ApJS*, **90**, 1
Eracleous, M. & Halpern, J., 2003, *ApJ*, **599**, 886
Halpern, J., 1990, *ApJ*, **365**, 51
Halpern, J. & Eracleous, M., 1994, *ApJ*, **433**, 17
Inoue, H. & Matsumoto, C., 2003, *PASJ*, **55**, 625
Iwasawa, K. *et al.*, 1996, *MNRAS*, **282**, 1038
Marziani, P. *et al.*, 2003, *ApJS*, **145**, 199
Miller, J. & Peterson, B., 1990, *ApJ*, **361**, 98
Misra, R., 2001, *MNRAS*, **320**, 445
Nandra, K. *et al.*, 1997a, *ApJ*, **476**, 70
Nandra, K. *et al.*, 1997b, *ApJ*, **477**, 602
Page, M. *et al.*, 2003, *MNRAS*, **343**, 1241
Perez, E. *et al.*, 1988, *MNRAS*, **230**, 353
Reynolds, C., 1997, *MNRAS*, **286**, 513
Storchi-Bergmann, T. *et al.*, 1997, *ApJ*, **489**, 87
Strateva, I. *et al.*, 2003, *AJ*, **126**, 1720
Sulentic, J. *et al.*, 1990, *ApJ*, **355**, L15
Sulentic, J. *et al.*, 1995, *ApJ*, **438**, L1
Sulentic, J. *et al.*, 1998a, *ApJ*, **497**, 65
Sulentic, J. *et al.*, 1998b, *ApJ*, **501**, 54
Sulentic, J. *et al.*, 1999, In *PASP Conf. #175*, 175
Tanaka, Y. *et al.*, 1995, *Nature*, **375**, 659
Vaughn, S. & Fabian, A., 2004, *MNRAS*, **348**, 1415
Wilms, J. *et al.*, 2001, *MNRAS*, **328**, 27
Zheng, W., Binette, L., & Sulentic, J., 1990, *ApJ*, **365**, 115
Zheng, W., Veilleux, S., & Grandi, S., 1991, *ApJ*, **418**, 196

Discussion

Q: B. LEMPEL:

In an expanding gas bubble (e.g., supernova . . .), the gas pressure tends to push the dust into a shell shape towards the outer part of the bubble. The velocities of gas and dust decrease from the inside to the outside of the bubble. An Hα line, for example, is split (at the edges of the bubble) with one side being redshifted and the other blueshifted. However, because of the cosmological redshift, the redshifted component will be in the IR while the blueshifted component will be in the visible. Due to this splitting of velocity, absorption due to dust can act as a broad-band filter, which absorbs all the visible light and, therefore, all the lines in the visible. The object can then appear as an "ultra cosmological" object as the IR radiation is not absorbed.

A: J. S.:

If I understand you correctly, the velocity separation in the expanding bubble that you describe would require an expansion velocity in excess of the speed of light and would therefore be impossible. In the visible an expansion velocity of 1/3c would require a separation of Hα peaks of $\sim$2200 angstroms or an FeKα profile width of about 3–4 keV in the X-ray.

Q: J. SUPERNANT:
Optical fiber spectrographs are now common. Has anyone tried using two optical fibers to "turn" around a QSO in order to resolve the two jets, or two sides of an accretion disk?

A: J. S.:
As also pointed out by Professor Pecker, we can not spatially resolve the two sides of an accretion disk (if it exists) – even from space. We are spectroscopically resolving something in quasars with double-peaked emission lines but we cannot agree what that "something" is (e.g., accretion disk, bicone/jet flow, binary black hole, or none of the above).

Comment: M. MOLES:
Evolution should have to take into account the need of a physically large volume to capture not only some average values but also the variance at every z. This is hard to do and very rarely done to low z-values. This is a real weakness for many of the claims of evolution seen in some parameters. Moreover, in many cases, the final results are dependent on the specific metric used. All in all, the quantitative arguments on evolution are not too many!

Q: H. ARP:
What is the situation with ejection? Do you see evidence in the double-peaked lines for ejection of material?

A: J. S.:
We have considered a bicone outflow model for the sources that show double-peaked emission lines. The model works better than standard accretion disk fits and accounts for their rarity in a natural way (preferred orientation). It is however not in fashion at this time. The recent paper by Eracleous & Halpern 2003 (cited in my text) argues against it.

Q: J. V. NARLIKAR:
I have been trying to get from the specialists on accretion disks, the size of the disk for a given black hole mass, say 10^9 M$_\odot$. I could not get a straight answer. Do you know the answer?

A: J. S.:

One can estimate the size of accretion disks theoretically or experimentally. The size of the putative disk in quasars is often modeled in the range of 10-25 000 Schwarzschild radii (where $r_g = GM/c^2$). The disk is thought to become gravitationally unstable at larger radii. If one answers in an empirical way then, if optical emission lines come from the disk, then reverberation mapping results suggest that the radiating part of the disk is 1–100 light days from the central continuum source.

Comment: A. BLANCHARD:

The integrated emissivity of galaxies is claimed to show an evolution of the average population of galaxies, increasing first rapidly and then flattening or decreasing at higher redshift.

Q: J. SURDEJ:

Mike Hawkins has produced one of the most complete surveys of quasars based upon photometric variability over long time scales. His work has been published. He was able to pick up all quasars identified by means of independent techniques (color, X-ray, etc.).

A: J. S.:

It is an interesting way to search for quasars. If I am not mistaken it involves only one field (ESO/SERC 287) with a long temporal baseline of observations. I am not aware of any discussion of this sample in the context of completeness.

Q: A. BLANCHARD:

There are few "standard cosmologists" who acknowledge that there is now a covariant formulation of the MOND-like approach that is in very acceptable agreement with observations.

A: J. S.:

If the number of people who acknowledged a result was proportional to the physical consistency underlying it then there would be 10^3 m for MOND and 10^{-3} m for accretion disks. That does not mean that MOND is correct or that accretion disks are non-existent. Acknowledgement is usually 90% ideology and 5% awareness of the facts!

Part II

Observational facts relating to background radiation

.

4

CMB observations and consequences

François R. Bouchet

Institut d'Astrophysique de Paris, CNRS, 98 bis Bd Arago, Paris, F-75014, France

Abstract

I describe briefly the Cosmic Microwave Background (hereafter CMB) physics, which explains why high accuracy observations of its spatial structure are a unique observational tool to test our cosmological models, determine the global cosmological parameters, and constrain observationally the physics of the early Universe. I also briefly survey the many experiments that have measured the anisotropies of the CMB and led to crucial advances in observational cosmology. The somewhat frantic series of new results has culminated in the outcome of the WMAP satellite, which confirmed earlier results, set new standards of accuracy, and suggested that the Universe may have reionized earlier than anticipated. Many more CMB experiments are currently taking data or being planned, offering opportunities to challenge further the current concordance model. The large increase in accuracy promises the possibility of falsifying or consolidating even more strongly the current paradigm, which has already met with considerable predictive success.

1 Introduction

As we shall see, the analysis of the CMB temperature anisotropies indicates that the total energy density of the Universe is quite close to the so-called critical density, ρ_c, or equivalently $\Omega = \rho/\rho_c \simeq 1$. We therefore live in a close-to-spatially-flat Universe. In agreement with the indications of other cosmological probes, the team of the CMB satellite WMAP [4] found that about 30% of that density appears to be contributed by matter ($\Omega_M = 0.29 \pm 0.07$), most of which is dark – i.e., not interacting electromagnetically – and cold – i.e., its primordial velocity dispersion can be neglected. The usual atoms (the baryons) contribute less than about 5% ($\Omega_B \simeq 0.047 \pm 0.006$) [18]. If present, a hot dark matter component does not play a significant role in determining the global evolution of the Universe. While many candidate particles have been proposed for this CDM, it has not yet been detected in laboratory experiments, although the sensitivities of the latter are now reaching the range in which realistic candidates may lie. The other $\sim 70\%$ of the critical density is contributed by a smoothly distributed and slowly varying "vacuum" energy density or dark energy, whose net effect is repulsive, i.e., it tends to accelerate the expansion of the Universe. Alternatively, this effect might arise from the presence in Einstein's equation of the famous cosmological constant term, Λ. While this global census, surprising as it may be, has already been around for some time (see Section 3), the WMAP results have tightened earlier constraints and gave further confidence to the model. These constraints obtained from the analysis of CMB anisotropies arise from – and confirm – the current theoretical understanding of the formation of large scale structures in the Universe, which we now briefly outline.

The spatial distribution of galaxies revealed the existence of the large scale structures (clusters of size ~ 5 Mpc, filaments connecting them, and voids of size ~ 50 Mpc), whose existence and statistical properties can be accounted for by the development of primordial fluctuations by gravitational instability. The current paradigm is that these fluctuations were generated in the very early Universe, probably during an inflationary period; that they evolved linearly over a long period, and more recently reached density contrasts high enough to form bound objects. Given the census given above, the dominant component that can cluster gravitationally is Cold Dark Matter (CDM).

The analysis of the CMB anisotropies also indicates that the initial fluctuations statistics had no large deviations from a Gaussian distribution and that they were mostly adiabatic, i.e., all components (CDM, baryons, photons) had the same spatial distribution. The power spectrum of the "initial conditions" appears to be closely approximated by a power law $P(k) = \langle |\delta_k|^2 \rangle = A_S k^{n_S}$, where δ_k stands for the Fourier transform of the density contrast ($P(k)$ is therefore the Fourier transform of the two-point spatial correlation function). The logarithmic slope, n_S, is quite close to unity ($n_S = 0.99 \pm 0.04$ from WMAP alone [18]). This shape implies that small scales collapsed first, followed by larger scales, with small objects merging to form bigger objects. The formation of structures thus appears to proceed *hierarchically* within a "cosmic web" of larger structures of contrast increasing with time.

Figure 4.1(a) shows the generated structures in the CDM components in a numerical simulation box of 150 Mpc, while Figure 4.1(b) shows the evolution with redshift of the density in a thin slice of that box. The statistical properties of the derived distribution (with the cosmological parameters given above) appear to provide a close match to those derived from large galaxy surveys. Note that this simulation with $\Omega_\Lambda = 2/3$, $\Omega_M = 1/3$ was performed in 2000, well before the WMAP results. Indeed it was already then the favorite model.

When collapsed objects are formed, the baryonic gas initially follows the infall. But shocks will heat that gas, which can later settle in a disk and cool, and form

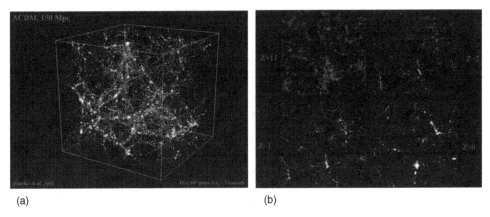

(a)   (b)

Figure 4.1 A numerical simulation in a 150 Mpc box of a LCDM Universe ($\Omega = 1$, $\Omega_\Lambda = 2/3$, $n_S = 1$). (a) Resulting distribution of the CDM at present (luminosity proportional to the density). (b) Temporal evolution by gravitational instability of a thin (15 Mpc) slice of the box showing the hierarchical development of structures within a global cosmic web of increasing contrast, but quite discernible early on.

stars and black holes, which can then feed back through ionizing photons, winds, supernovae . . . on the evolution of the remaining gas (after first reionizing the Universe at $z > 6$). In this picture, galaxies are therefore (possibly biased) tracers of the underlying large scale structures of the dark matter.

2 Physics overview of CMB anisotropies

In this standard cosmological model, processes in the very early universe generate the seed fluctuations that ultimately give rise to all the structures we see today. In the early Universe, baryons and photons were tightly coupled through Thomson scatterings of photons by free electrons (and nuclei equilibrated collisionally with electrons). When the temperature in the Universe became smaller than about 3000 K (which is much lower than 13.6 eV due to the large number of photons per baryon $\sim 1.5 \times 10^9$), the cosmic plasma recombined and the ionization rate x_e fell from 1 at $z > 1100$ down to $x_e < 10^{-3}$ at $z < 1100$: The photons mean free path $\propto 1/x_e$ rapidly became much larger than the horizon $\sim cH^{-1}$. As a result, the Universe became transparent to background photons over a narrow redshift range of 200 or less. Photons then propagated freely as long as galaxies and quasars did not reionize the Universe (but by then the density would have fallen enough that only a small fraction was rescattered). We therefore observe a thin shell around us, the last scattering "surface" (LSS in short) where the overwhelming majority of photons last interacted with baryonic matter, at a redshift of 1100, when the Universe was less than 400 000 years old. The anisotropies of the CMB are therefore the imprint

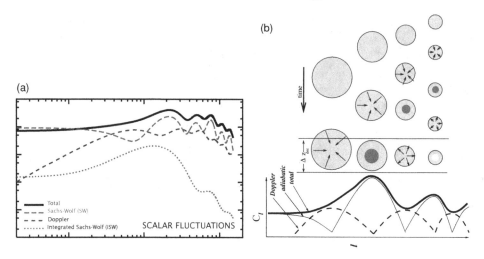

Figure 4.2 The expected shape of the angular power spectrum of the temperature anisotropies, $C(\ell)$ (times ℓ^2 to give the logarithmic contribution of each scale to the variance). (a) Relative contributions; it has been assumed that only scalar fluctuations are present. The plot is in log–log coordinates. (b) As time progresses, larger and larger fluctuations start oscillating and leave their characteristic imprint on the spectrum (reprinted from [13]).

of the fluctuations as they were at that time (but for a small correction due to the photons' propagation through the developing Large Scale Structures).

To analyze the statistical properties of the temperature anisotropies, we can either compute the angular correlation function of the temperature contrast δ_T, or the angular power spectrum $C(\ell)$, which is its spherical harmonics transform (in practice, one transforms the δ_T pattern in $a(\ell, m)$ modes and sums over m at each multipole since the pattern should be isotropic, at least for the trivial topology, in the absence of noise). A given multipole corresponds to an angular scale $\theta \sim 180°/\ell$. These two-point statistics completely characterize a Gaussian field. Figure 4.2(a) shows the expected $C(\ell)$ shape in the context we have described above.

This specific shape of the $C(\ell)$ arises from the interplay of several phenomena. The most important is the so-called "*Sachs–Wolfe effect*" [16], which is the energy loss of photons that must "climb out" of gravitational potential wells at the LSS (ultimately to reach us to be observed), an effect that superimposes on the intrinsic temperature fluctuations (we therefore observe an effective temperature that sums this effect, of opposite sign for adiabatic initial conditions where every component [γ, baryons, CDM ...] is perturbed simultaneously). Figure 4.2(b) gives a pictorial view of the temporal development of primordial fluctuations at different scales (top) and how the state of fluctuations translates at recombination in the power spectrum of CMB fluctuations (bottom).

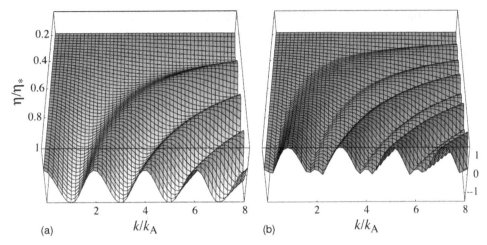

Figure 4.3 Temporal evolution of the effective temperature, which sums the effects (of opposite sign) of the intrinsic temperature and of the Newtonian potential fluctuations (for $R = (p_B + \rho_B)/(p_\gamma + \rho_\gamma) = cste$). (a) Amplitude. Note the zero point displacement, which leads to a relative enhancement of compressions. (b) rms showing the enhanced odd-numbered peaks. Reprinted from [10].

Since the density contrasts of these (scalar) fluctuations is very weak, one can perform a linear analysis and study each Fourier mode independently (the effect of the primordial spectrum will thus simply be to weight the various modes in the final $C(\ell)$). Figure 4.3 shows the (approximate) temporal evolution of the amplitude of different Fourier modes. Gravity tends to enhance the contrast, the (mostly photonic) pressure resists and at some points stops the collapse that bounces back, and expands before recollapsing. This leads to acoustic oscillations, on scales small enough that the pressure can be effective, i.e., for $k > k_A$, where the acoustic scale k_A is set by the inverse of the distance traveled at the speed sound of at the time $\eta_\star$ considered. On scales larger than the sound horizon ($k < k_A \propto 1/(c_S\,\eta_\star)$), the initial contrast is simply amplified. At $k = k_A$ the amplification is maximal, while at $k = 2k_A$ it has time to bounce back fully. More generally, the odd-multiples of k_A are at maximal compression, while this is the opposite for the even multiples of k_A.

One should note the displacement of the zero point of the oscillations, which results from the inertia that baryons bring to the fluid. The rms of the modes amplitude (right plot) therefore show a relative enhancement of the odd (compression) peaks versus the even (rarefaction) ones, this enhancement being directly proportional to the quantity of baryons, i.e., $\Omega_B h^2$, where h stands for the Hubble "constant," $H = \dot{a}/a$, in units of 100 km s^{-1} Mpc^{-1} (today $h_0 = 0.72 \pm 0.05$). Note that since Ω_X stands for the ratio of the density of X to the critical density (such that the Universe is spatially flat), and since that critical density decreases with time as h^{-2}, $\Omega_X h^2$ is indeed proportional to the physical density of X.

Let us assume that the LSS transition from opaque to transparent is instantaneous, at $\eta = \eta_\star$. What we would see then should just be the direct image of these standing waves on the LSS; one therefore expects a series of peaks at multipoles $\ell_A = k_A \times D_\star$, where $D_\star$ is the angular distance to the LSS, which depends on the geometry of the space-time. Of course a given k contributes to some range in ℓ (when k is perpendicular to the line of sight, it contributes to lower ℓ than when it is not), but this smearing is rather limited. The dependence of the acoustic angular scale ℓ_A on geometry and speed of sound leads to its dependence on the values of three cosmological parameters. One finds for instance

$$\frac{\Delta\ell_A}{\ell_A} \simeq -1.1\frac{\Delta\Omega}{\Omega} - 0.24\frac{\Delta\Omega_M h^2}{\Omega_M h^2} + 0.07\frac{\Delta\Omega_B h^2}{\Omega_B h^2} \tag{1}$$

around a flat model $\Omega = 1$ with 15% of matter ($\Omega_M h^2 = 0.15$) and 2% of baryons [10]. Note that this information on the peaks' positions (and in particular that of the first one) is mostly dependent on the total value of Ω (geometry), with some weaker dependence on the matter content $\Omega_M h^2$, and an even weaker one on the baryonic density.

Concerning the latter, as already mentioned, baryons increase the inertia of the baryon–photon fluid and shifts the zero point of the oscillations. A larger baryonic density tends to increase the contrast between odd and even peaks; one can therefore use this contrast-in-height information as a baryometer. The influence of dark matter is more indirect. By increasing its quantity, one increases the total matter density and advances the time when matter comes to dominate the energy density of the Universe. This changes the duration of time spent in the radiation and matter dominated phase, which may have different growth rates. The net effect is to decrease globally the first peak's amplitude when the matter content increases (in addition to the small shift in scale due to the variation of ℓ_A already noted above). The effect on the power spectrum peaks of all the matter is thus rather different from that of the baryonic component alone. Therefore the shape of the spectrum is sensitive to both separately. This suggests that degeneracies in the effect of these three parameters (Ω, $\Omega_M h^2$, $\Omega_B h^2$) can be lifted with sufficiently accurate CMB measurements, a statement that more detailed analyses confirm.

The reader in a hurry can skip to the conclusion of this section as I now turn to describing other effects that must be taken into account to understand fully the shape of the power spectrum. Since the fluid is oscillating, there is also a *Doppler effect* in the k direction (dashed line in Figure 4.2(a)), which is zero at the acoustic peaks and maximal in between. This effect adds in quadrature to the Sachs–Wolfe effect considered so far. Indeed, imagine an acoustic wave with k

perpendicular to the line of sight, we see no Doppler effect, while for a k parallel to the line of sight, the Doppler effect is maximal and the Sachs–Wolfe effect is null. This smoothes out the peak and trough structure, although not completely since the Doppler effect is somewhat weaker than the SW effect (by an amount $\propto \Omega_B h^2$). Therefore an increase in the baryonic abundance also increases the peak–trough contrast (in addition to the odd–even peak contrast).

So far we considered the fluid as perfect and the transition to transparence as instantaneous, none of which is exactly true. Photons scattered by electrons through Thomson scattering in the baryons–photons fluid perform a random walk and diffuse away proportionally to the square root of the time (in comoving coordinates, which remove the effect of expansion). Being much more numerous than the electrons by a factor of a few billion, they drag the electrons with them (which by collisions in turn drag the protons). Therefore all fluctuations smaller than the diffusion scale are severely damped. This so-called Silk damping is enhanced by the rapid increase of this diffusion scale during the rapid but not instantaneous combination of electrons and protons that leads to the transparence. As a result of the finite thickness of the LSS and the imperfection of the fluid, there is *an exponential cut-off of the large-ℓ part* of the angular power spectrum. As a result, there is not much primordial pattern to observe at scales smaller than $\sim 5'$.

After recombination, photons must travel through the developing large scale structures to reach the observer. They can lose energy by having to climb out of potential wells that are deeper than when they fell in (depending on the rate of growth of structures, which depends in turn on the cosmological census). Of course the reciprocal is also true, i.e., they can gain energy from forming voids. These tend to cancel at small scale since the observer only sees the integrated effect along the line of sight. The dotted line of Figure 4.2 shows the typical shape of that *Integrated Sachs-Wolfe* (ISW) contribution. The ISW is anti-correlated with the Sachs-Wolfe effect, so that the total power spectrum $C(\ell)$ is in fact a bit smaller than the sum of each spectrum taken separately. Finally, other small secondary fluctuations might also leave their imprint, like the lensing of the LSS pattern by the intervening structures, which slightly smoothes the spectrum. But that can be predicted accurately too. And in fact the smoothing kernel dependence on cosmological parameters introduces small effects that may help in reducing some residual degeneracies between the effect of parameters on the power spectrum shape.

Other secondary effects, imprinted after recombination, are generally much weaker (at scales $> 5'$). For instance, the *Rees–Sciama* effect [15] (a non-linear version of the ISW) generates temperature fluctuations, with amplitudes of about a few 10^{-7} to 10^{-6}; its amplitude is maximum for scales between 10 and 40 arc minutes [17]. At the degree scale, this contribution is only of the order of 0.01 to 0.1%

of the primary CMB power. The inverse Compton scattering of the CMB photons on the free electrons of hot intra-cluster gas produces the *Sunyaev and Zel'dovich (SZ) effect* [19, 20]). This effect has a specific spectral signature that should allow it to be separated, at least in sufficiently sensitive multi-frequency experiments. But the motion along the line of sight of clusters induces a first order Doppler effect, usually called the *Kinetic Sunyaev–Zel'dovich* effect, which is a true source of temperature fluctuation, albeit rather weak (the rms cluster velocity is $\sim 10^{-3}c$) and in the specific direction of clusters. A similar effect, the *Ostriker–Vishniac effect* [14, 22], arises from the correlations of the density and velocity perturbations along the line of sight, when the Universe is totally ionized. The corresponding anisotropies are at the few-arc-minute scale and their amplitudes depend much on the ionization history of the Universe [7, 11]). However, they remain smaller than the primary anisotropies for $\ell < 2000$. This type of Doppler effect can in fact happen in all sorts of objects containing ionized gas, such as expanding shells around the sources that reionized the Universe [2, 9, 12, 13]), or primordial galaxies hosting super-massive black holes [1], but the relevant angular scales are rather in the arc second range or smaller. This does not hold true however for various foreground emissions, such as those of our own Galaxy. But as for the Sunyaev-Zel'dovich effect, one can use multi-frequency observations to separate them out rather well.

In summary, *the seeds of large scale structures must have left an imprint on the CMB, and the statistical characteristics of that imprint can be precisely predicted as a function of the properties of the primordial fluctuations and of the homogeneous Universe.* Reciprocally, we can use measurements of the anisotropies to constrain those properties.

3 Observations of CMB anisotropies

The first clear detection of CMB anisotropies was made in 1992 by the DMR experiment aboard the COBE satellite orbiting the Earth with the DMR instrument (and soon afterwards by FIRS), with a ten degree (effective) beam and a signal to noise per pixel of around 1. This led to a clear detection of the large scale, low-ℓ, Sachs–Wolfe effect, the flatness of the curve (see Fig. 4.4(a)) indicating that the logarithmic slope of the primordial power spectrum, n_S, could not be far from 1. The $\sim 30 \, \mu K$ height of the plateau gave a direct estimate of the normalization of the spectrum, A_S (assuming the simplest theoretical framework, without any possible direct checks of the other predictions given the data).

In the next four years (Fig 4.4(a)), a number of experiments started to suggest an increase of power around the degree scale, i.e., at $\ell \sim 200$. As shown by Fig. 4.4(b), by 1999 there was a clear indication from many experiments taken

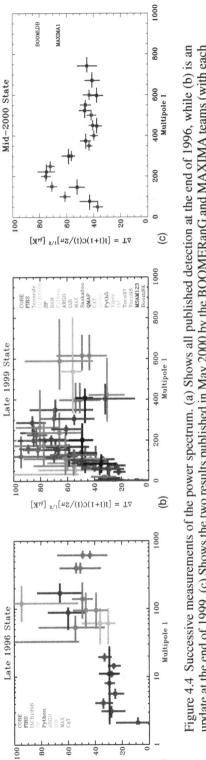

Figure 4.4 Successive measurements of the power spectrum. (a) Shows all published detection at the end of 1996, while (b) is an update at the end of 1999. (c) Shows the two results published in May 2000 by the BOOMERanG and MAXIMA teams (with each curve moved by + or −1 σ of its respective calibration).

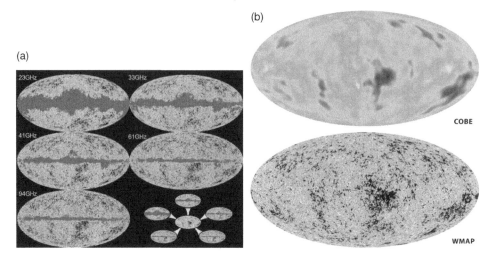

Figure 4.5 (a) The WMAP maps at each frequency, which shows in particular the varying strength of the galactic emissions according to frequency, and in the bottom right an icon illustrating that this information is merged to extract an estimate of the CMB part. (b) The deduced map of the anisotropies compared with the previous one from the COBE satellite. Source NASA/WMAP Science Team.

together that a first peak had been detected. But neither the height nor the location of that peak could be determined precisely, in particular in view of the relative calibration uncertainties (and possible residual systematics errors).

That situation changed in May 2000 when the BOOMERanG and Maxima collaborations both announced a rather precise detection of the power spectrum from $\ell \sim 50$ to $\ell \sim 600$. That brought a clear determination of the first peak around an ℓ of 220 (see panel (c)), with the immediate implications that Ω had to be close to 1. This result had considerable resonance since it clearly indicated, after decades of intensive work, that the spatial geometry of the Universe is close to flat, with of course the imprecision due to the poor determination of the other parameters, which also have an influence, albeit weaker, on the position of that peak (see Equation (1)).

As recalled earlier, a crucial prediction of the simplest adiabatic scenario is the existence of a series of acoustic peaks whose relative contrast between the odd and even ones gives a rather direct handle on the baryonic abundance. In addition, one expects to see the damping tail at larger ℓ. All of these have now been established by the DASI 2001 experiment, an improved analysis of BOOMERanG, and in particular by the release in May of 2002 of the VSA and CBI results. In addition the Archeops experiment at the end of gave 2002 a quite precise determination of the low-ℓ part of the spectrum. Panel (c) of Fig. 4.4 shows a co-analysis performed by

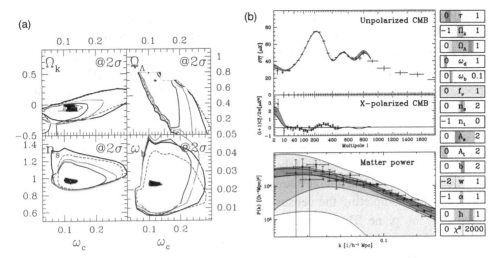

Figure 4.6 (a) Shows the successive constraints in the $\Omega_k = 1 - \Omega$, Ω_Λ, n_S, $\omega_B = \Omega_B h^2$ versus $\omega_c = \Omega_{CDM} h^2$ cuts in the global parameter space fitted to the $C(\ell)$ successive data, using COBE/DMR in all cases. The line coding is the following: CBI = black, VSA = (outer) dotted, Archeops = light gray, ACBAR = dark gray, Ruhl cut for BOOMERanG = mid gray. The gray shaded area in the top left panel corresponds to Acbar + Archeops + Ruhl + DASI + Maxima + VSA + CBI. The interior dotted line is all of the above + WMAP with no prior on τ, while black is with a τ prior motivated by their "model independent" result from the TE analysis (it is broader than a 0.16 ± 0.04 Gaussian) (Courtesy D. Bond). (b) Is explained in the text. Reproduced from [21].

D. Bond of all results obtained up to the end of 2002, as well as the recent determination by the WMAP satellite. Clearly all the pre-WMAP ground experiments had done quite a wonderful job at pinning down the shape of the temperature spectra. This panel also shows the spectrum as we know it today, when all experimental results are analyzed together.

Figure 4.6(a) shows the constraints successively posed by these CMB experiments on some of the parameters of the model, using only weak bounds arising from other cosmological studies. These bounds state that the current Hubble "constant," $H_0 = 100h^{-1}$ km s^{-1} Mpc^{-1}, has to have a value between 45 and 90 km s^{-1} Mpc^{-1}, that the age of the Universe has to be greater than 10 billion years, and that the matter density is larger than 1/10 of the critical density, all of which can be considered as very well established (if for instance the Universe has to be older than its oldest stars!).

The top left panel shows that indeed the curvature term $\Omega_k = 1 - \Omega$ has to be close to zero. The panel on the right shows that Ω_Λ and $\omega_c = \Omega_{CDM} h^2$ are not well determined independently of each other by single experiments. This simply reflects the fact that the $C(\ell)$ global pattern scales by the angular distance (recall $\ell_A = k_A D_\star$), which is determined by the geometry (i.e., $\Omega = \Omega_\Lambda + \Omega_{CDM} (+\Omega_B)$), while the data are not precise enough to uncover the subtler effects (sound speed, lensing . . .) that break that degeneracy. But this degeneracy was lifted by the co-analysis, even before WMAP, and independently of the supernovae result . . .

The bottom left panel lends support to the $n_s = 1$ hypothesis. Many inflationary models suggest values of n_s slightly lower than one (and even departures from a pure power law), but the data are not yet good enough to address these questions convincingly. Completing the census, the bottom right panel shows the contours in the $\omega_b - \omega_c$ plane. The CMB determination turns out to be in excellent agreement with the constraints from primordial nucleosynthesis calculations, which yield $\omega_b = \Omega_B h^2 = 0.019 \pm 0.002$.

Figure 4.6 (b) displays the current determination (points with error bars) of the power spectrum of CMB temperature anisotropies from WMAP and other experiments (top), that of the cross-correlation between the temperature and the (E-part of) polarization from WMAP alone), while the data points at the bottom give the matter power spectrum measured around us with the Sloan Digital Sky Survey. This plot therefore summarizes our current statistical knowledge of fluctuations, at times separated by about thirteen billion years. The light gray, mid gray, and dark gray regions show the result of a fit *on the CMB data alone* using, respectively a 12, 10, 7 parameter model. In all cases, the predicted range for the matter power spectrum (measured about 13 billions years later) is in excellent agreement with the *prediction* from the CMB. The minimal 7 parameters model includes $(\tau, \Omega_\Lambda, \omega_B, \omega_c, A_S, n_s, b)$ (b accounts for a possible biasing of light versus mass). The model with 10 parameters allows in addition for a non-flat Universe ($\Omega_{tot} \neq 1$), with a tensorial contribution ($A_T \neq 0$), and a running spectral index (i.e., $d \ln n_s / d \ln k \neq 0$), while the maximal 12 parameter model further allows for a non-zero contribution from massive neutrinos, as well as a vacuum energy with a non-standard equation of state ($p = w\rho$, with $w \neq -1$). This shows that there is no need in the data for a more exotic model than the current, minimal concordance model with 7 parameters. Since the CMB and the large scale distribution of galaxies agree, we can additionally co-analyze these spectra, which yields the blue region. It turns out that the corresponding values of the parameters, which are rather tightly constrained, agree with the inference on the baryonic abundance from nucleosynthesis, the value of the Hubble constant derived by the Hubble key project, and the total mass density inferred for instance from gravitational lensing studies. This network of independent evidence giving a consistent view is rather convincing.

In summary, many of the theoretical predictions corresponding to the simplest scenario for the generation of initial conditions (Gaussian statistics, adiabatic modes, negligible tensorial contribution, a scale invariant power spectrum) in a flat Universe dominated by dark energy and cold dark matter have now been detected, from the Sachs–Wolfe plateau, to the series of peaks starting at $\ell \simeq 220$, to the damping tail, together with a first detection of the CMB polarization at the expected level. The derived parameters are consistent with the various constraints from other cosmological probes and there are no glaring signs of inconsistencies. This was in fact in place before the WMAP results, but it is remarkable that adding WMAP essentially zooms in onto the expected values, bringing now further support to the model and its parameters, as recalled in Section 1.

4 Perspectives

Cosmology therefore has a great *concordance* model, based on a minimal 7 (8) parameters set $(\Omega, \Omega_{CDM}, \Omega_B, H, \tau, (b)$ to describe the global evolution, and n_S, A_S to describe the initial conditions, b to describe a possible light versus mass bias), which fits quite well all (multiple and partially independent) observational evidence. But it might be too early to consider that we have definitively established the *standard* cosmological model, corroborated by *many* independent probes.

In order to check the strength of the edifice, let us for instance consider what it takes to stick to a simpler Einstein–de Sitter model (EdS, with $\Omega_{Matter} = 1, \Omega_\Lambda = 0$) with no dark energy. As a matter of fact, Blanchard *et al.* find in [5] a great fit to the currently measured $C(\ell)$ and a quite reasonable one to the $P(k)$, provided they assume that (1) initial conditions are not scale invariant (the authors consider an initial spectrum with 2 slopes) and (2) there exists a non-clustering component $\Omega_x = 0.12$, both of which are quite possible. The supernovae data would then be the only source of independent evidence for a non-zero Ω_Λ, and many would question accepting the existence of such a theoretically unsettling component on that basis. But this particular Einstein–de Sitter model does require also that $H_0 \sim 46$ km s^{-1} Mpc^{-1}. This would require in turn that the HST measurement of $H_0 = 72 \pm 8$ km s^{-1} Mpc^{-1} be completely dominated by an as-yet unknown systematics effect, which most cosmologists are not ready to accept easily. Accepting this model would thus require discarding two independent lines of evidence.

In addition we may now have some rather direct evidence of the presence of dark energy (meaning here $\Omega_\Lambda \neq 0$). Indeed, Bough and Crittenden [5] and Fosalba and Gaztanaga [7] found a significant cross-correlation between the CMB anisotropies measured by WMAP and various tracers of Large Scale Structures. This was anticipated since the evolution of potential wells associated to developing Large Scale Structures when CMB photons travel through them generally leaves an imprint, which we have already described, the Integrated Sachs–Wolfe effect (ISW). This imprint must of course be correlated with tracers of LSS. It is interesting to note that, in the EdS model, gravitational potential wells are linearly conserved in the matter era, in which case no correlation should be found. Instead Bough and Crittenden [5] found a non-zero value at the 2.4 to 2.8 σ level at zero lag of the cross-correlation function of the WMAP map data with the hard X-ray background measured by the HEAO-1 satellite. They also found a somewhat less significant correlation (at the 1.8 to 2.3 σ level) with an independent tracer, the radio counts from the NVSS catalog. Fosalba and Gaztanaga [7] used instead the APM galaxy catalog to build the (projected) density field by smoothing at 5.0° and 0.7° resolution and they also found a substantial cross-correlation ... While these detections may not yet be at a satisfactory level of significance for such an important implication, it does bring a third line of evidence against the model proposed in [5].

Let us therefore assume that this concordance model offers at least a good first-order description of the Universe. Still, deviations from this minimal description remain quite possible and interesting. One possibility much debated recently is that what appears to be the manifestation of a cosmological constant is rather a dynamical entity, for instance a quintessence field with an equation of state where the pressure to density ratio is equal to $w(z)$ (the cosmological constant corresponding to the case $w = -1$). Another possibility concerns a small contribution from massive neutrinos. In both cases better CMB data might help determine these effects (as well as many others, not even mentionned so far, like an isocurvature initial contribution, or one from various topological defects, or deviations from standard gravity). The domain where most progress is expected and eagerly awaited concerns the characterization of initial conditions and its implications for physics of the early Universe.

5 Conclusions

Measurements of the CMB are unique in the ensemble of astrophysical observations that are used to constrain cosmological models. They have the same character as fundamental physics experiments; they relate fundamental physical parameters describing our world to well-specified signatures, which can be predicted beforehand with great accuracy.

The knowledge of CMB anisotropies has literally exploded in the last decade, since their momentous discovery in 1992 by the DMR experiment on the COBE satellite. Since then, the global shape of the spectrum has been uncovered thanks to many ground and balloon experiments and most recently the WMAP satellite, so far confirming the simplest inflationary model and helping shape our surprising view of the Universe: spatially flat, and dominated by dark energy (or Λ) and cold dark matter (25%), with only a few per cent of atoms. But the quest is far from over, with many predictions still waiting to be checked and many parameters in need of better determination.

If the next 10 years are as fruitful as the last decade, many cosmological questions should be settled, from a precise determination of all cosmological parameters to characteristics of the mechanism that seeded the growth of structures in our Universe, if something even more exciting than what is currently foreseen does not emerge from the future data . . .

References

[1] N. Aghanim, C. Balland, and J. Silk. Sunyaev-Zel'dovich constraints from black hole-seeded proto-galaxies. *A&A*, **357**:1–6, May 2000.

[2] N. Aghanim, F. X. Desert, J. L. Puget, and R. Gispert. Ionization by early quasars and cosmic microwave background anisotropies. *A&A*, **311**:1–11, July 1996.

[3] C. L. Bennett, M. Halpern, G. Hinshaw, N. Jarosik, A. Kogut, M. Limon, S. S. Meyer, L. Page, D. N. Spergel, G. S. Tucker, E. Wollack, E. L. Wright, C. Barnes, M. R. Greason, R. S. Hill, E. Komatsu, M. R. Nolta, N. Odegard, H. V. Peiris, L. Verde, and J. L. Weiland. First-year Wilkinson Microwave Anisotropy Probe (WMAP) observations: Preliminary maps and basic results. *ApJS*, **148**:1–27, Sept. 2003.

[4] A. Blanchard, M. Douspis, M. Rowan-Robinson, and S. Sarkar. An alternative to the cosmological "concordance model." *A&A*, **412**:35–44, Dec. 2003.

[5] S. Boughn and R. Crittenden. A correlation of the cosmic microwave sky with large scale structure. *astroph/0305001*, 2003.

[6] S. Dodelson and J. M. Jubas. Reionization and its imprint of the cosmic microwave background. *ApJ*, **439**:503–516, Feb. 1995.

[7] P. Fosalba and E. Gaztanaga. Measurement of the gravitational potential evolution from the cross-correlation between WMAP and the APM Galaxy survey. *astroph/0305468*, 2003.

[8] A. Gruzinov and W. Hu. Secondary cosmic microwave background anisotropies in a universe reionized in patches. *ApJ*, **508**:435–439, Dec. 1998.

[9] W. Hu. CMB temperature and polarization anisotropy fundamentals. *Ann. Phys.*, **303**:203–225, Jan. 2003.

[10] W. Hu and M. White. CMB anisotropies in the weak coupling limit. *A&A*, **315**:33–39, Nov. 1996.

[11] L. Knox, R. Scoccimarro, and S. Dodelson. The impact of inhomogeneous reionization on cosmic microwave background anisotropy. *Phys. Rev. Lett.*, **81**:2004–2007, 1998.

[12] C. Lineweaver. Inflation and the cosmic microwave background. *Proceedings of the New Cosmology Summer School edt. M. Colless,* World Scientific, *astroph/0302213*, 2003.

[13] J. P. Ostriker and E. T. Vishniac. Generation of microwave background fluctuations from nonlinear perturbations at the era of galaxy formation. *ApJ*, **306**:L51–L54, July 1986.

[14] M. J. Rees and D. W. Sciama. Large scale density inhomogeneities in the universe. *ApJ*, **217**:511+, 1968.

[15] R. K. Sachs and A. M. Wolfe. Perturbations of a cosmological model and angular variations of the microwave background. *ApJ*, **147**:73, Jan. 1967.

[16] U. Seljak. Rees–Sciama effect in a cold dark matter universe. *ApJ*, **460**:549, Apr. 1996.

[17] D. N. Spergel, L. Verde, H. V. Peiris, E. Komatsu, M. R. Nolta, C. L. Bennett, M. Halpen, G. Hinshaw, N. Jarosik, A. Kogut, M. Limon, S. S. Meyer, L. Page, G. S. Tucker, and E. Wollack. First year Wilkinson Microwave Anisotropy Probe (WMAP) observations: Determination of cosmological parameters. *ApJ*, in press, *astroph/ 0302209*, 2003.

[18] R. A. Sunyaev and I. B. Zel'dovich. The velocity of clusters of galaxies relative to the microwave background – the possibility of its measurement. *MNRAS*, **190**:413–420, Feb. 1980.

[19] R. A. Sunyaev and Y. B. Zel'dovich. Formation of clusters of galaxies; protocluster fragmentation and intergalactic gas heating. *A&A*, **20**:189, Aug. 1972.

[20] M. Tegmark *et al.* Cosmological parameters from SDSS and WMAP. *Phys. Rev. D.*, **69(10)**:103501, May 2004.

[21] E. T. Vishniac. Reionization and small-scale fluctuations in the microwave background. *ApJ*, **322**:597–604, Nov. 1987.

[22] Y. B. Zel'dovich and R. A. Sunyaev. The interaction of matter and radiation in a hot-model universe. *Ap&SS*, **4**:301, 1969.

Discussion

Q: J. SULENTIC:

Are the 2–3 sigma (possible) correlations between the CMB fluctuations and radio, X-ray, and galaxy clusters related? Are they a potential problem for the concordance model? Assuming the answer to the last question is "no" would this become a problem if the significance rose to 5+ sigma and they could be attributed to a specific class or classes of low-z sources?

A: F. B.:

The propagation of CMB photons through the evolving gravitational potential of the large scale matter distribution is generically expected to leave an imprint, since the potential wells of a forming condensation might be deeper when the photon has to get out of it than when it fell in. One thus expects a redshift for photons emerging of such condensations and, reciprocally, a blueshift is anticipated for photons going through very large developing voids. The ensuing correlation of the CMB anisotropies with the density field revealed by any population tracing the large scale matter distribution (radio, X-ray, galaxy clusters, galaxies, etc.) has long been *predicted* (the one exception is for a flat Universe with only matter, a case currently excluded). As I mentioned in my talk, recent analyses do find such a correlation with various populations, *at a level consistent with expectations*, albeit at a rather low statistical significance due to the limitations of the current data (but see below, the answer to F. Bernardeau). Therefore these detections are currently rather a reinforcement of the concordance model than a problem. For the second part of the question, see the answer below to A. Blanchard.

Q: A. BLANCHARD:

Maybe a similar question would be what if you find a significant correlation, which is not at the level expected from the ISW in a concordance model?

A: F. B.:

Then three possibilities arise: (1) the data are misunderstood (or subject to unde-tected/unforeseen systematic effect), (2) we do not understand how the density field of the "tracer" population is related to the underlying matter field, or (3) there is something wrong in the model. Taking care of the first case (i.e., excluding that pos-sibility) amounts to accumulating many well-controlled observational probes, till the case is rock solid. In the second case, the "culprit" is, as was so many times the case, the complicated astrophysics that is needed to relate the mass distribution with that of light, and the art of cosmology is to try finding those crucial probes where the link is simplest (like CMB or lensing observation). In the third case, either this can be fixed by simple variations on the model (varying somewhat the values of the parameters, introducing a minor extra component, like a small fraction of hot dark matter), or we found a substantial flaw. An example would be to have a relationship between the physics of dark matter and dark energy resulting in a hidden relation that must be obeyed but is not enforced in the analysis (enforcing it might change the conclusions of our adjustments). It might also be even more exciting, like the smoking gun of totally new physics (e.g., deviations from general relativity). All cosmologists are hoping to find definite evidence for physics beyond the current model. But past history has shown that before doing that, one must spend a lot of energy on studying the more conventional explanations (misunderstood data, complicated astrophysical interpretation) before claiming a revolution is needed. Still the quality of the data and analyses has so much improved that it might be around the corner.

Q: F. BERNARDEAU:

Are there any hopes of improving the significance of the LSS-CMB correlations due to the ISW effect? With larger or deeper galaxy surveys, or with CMB polarization?

A: F. B.:

Yes, but this will neither be easy, nor fast (NB: LSS = Large Scale Structures). The ISW signal is at large scale, $\theta > 2°$ or $l < 100$. It arises from the contribution of different redshift ranges. To be more specific, in the concordance model, one finds that the dominant contributions come from the (observationally rather large) redshift range [0, 1.5], with a maximum around 0.5 (when dark energy becomes a sizable effect in current models). Note that on such large scales, the current CMB data from WMAP are essentially noise-free, and cover nearly all the sky. Thus no progress is to be anticipated from improvements in the measurements of the temperature field of the CMB. On the other hand, low-z surveys can be vastly improved. For a Lambda CDM (concordance) model, a perfect survey could yield a signal-to-noise ratio S/N = 7.5 (S/N = Σ $C_{tT}(l)^2/\Delta C_{tT}(l)^2$, where $C_{tT}(l)$ is the angular correlation function of the tracer field t with the temperature field T). Note that $\Delta C_{tT}(l)^2$ is inversely proportional to the fraction of the sky f_{sky} where the correlation can be computed, i.e., S/N $\propto f_{sky}$. If we think of the best present galaxy survey, the SDSS covers "only" of the sky, in a effective depth z of only ~ 0.1, with non-negligible Poisson sampling noise, which, as I showed, results in a S/N ~ 3. It will not be easy to do much better soon, but it is doable and will probably be done. Indeed an already rather ambitious all-sky galaxy survey with about 10 million galaxies uniformly distributed between $z = 0$ and 1 will yield an S/N of about 5 (see e.g., astroph/0401166). For polarization, the situation is different, since the CMB signal is much smaller, and we do not yet have any map. But Planck should allow having a sufficiently low noise on the E field determination at large scales to much reduce the limitation coming from the errors on the CMB, therefore providing the possibility of an independent check.

Q: D. ROSCOE:
The mass densities at the centers of galaxies are poorly predicted by theories involving dark matter. Suppose that the estimated mass densities in galactic centers were used as tight constraints, what would then be the effect on the concordance model?

A: F. B.:
If it turns out that modifications are indeed required by the data, there is a very wide range of possible explanations (see the general answer to A. Blanchard), and foremost that baryonic physics is not fully understood. One further possibility is to suppose minor modifications of the properties of the dark matter, which can alter the density profiles at small scale (e.g., auto-interactions) without altering any of the success of the concordance model at larger scales (CMB, lensing, Large Scale Structures, nucleosynthesis, etc.). Future tasks in cosmology are precisely to see whether we can convincingly exclude such possibilities and deduce that modification of the concordance model or the theory of gravity is requested from the data.

Q: J.-C. PECKER:

(1) Is not H a very badly defined quantity? i) It is basically an "average" over the distance d. H is indeed $\langle H \rangle_o{}^d$ ii) We have shown earlier that $\langle H \rangle$ indeed is decreasing with d from 100 or so to 50 or so.

(2) You derived *four* components of the fluctuations of the CMB, as observed at *five* frequencies. How unique is that decomposition in four components? What I would like to see is the spectrum $f(\lambda)$ for each of the pixels of the image. In other terms, how well are the five images calibrated in intensity?

A: F. B.:

(1) Of course H needs be defined by an average over a sufficiently large volume that we can rest assured that local measurements are not simple (local) statistical fluctuations.[1] The recent supernovae measurements show an amazingly tight linear relation of the distance modulus with redshift till[2] $z \sim 0.1$ ($v \sim 30\,000$ km s^{-1} Mpc^{-1}, and thus a scale $\sim 300\, h^{-1}$ Mpc). The largest voids in the *galaxy* distribution are rather at the $\sim 100\, h^{-1}$ Mpc, which strongly suggest we now probe scales sufficiently large that we have a (statistically) fair measurement. It is interesting to note that these local measurements are in good agreement with the results from model adjustment on the CMB measurements.

(2) The issue of component separation (of the CMB from foreground emission) is indeed very central to all CMB analysis. From that point of view, it is interesting to note that Archeops finds a consistent CMB sky with that of WMAP, but at a higher frequency.[3] Still, there is much debate in the CMB community on the issue of the *accuracy* of the process, i.e., what are the residuals from the separation (and how that propagates in the final error budget on, say, cosmological parameters). I personally would not be surprised if some of the precise numbers given by the WMAP team (including in particular the optical depth τ) turn out to be substantially revised after more careful work is performed (involving more templates obtained at other frequencies). But I do not expect radically new conclusions (except maybe for the value of τ). The general scientific community had the WMAP map available for about a year, and many people are currently working on that very topic in an attempt to challenge the WMAP team conclusions. Time will tell. Parenthetically, the Planck satellite that we are building (to be launched in 2007) will have *nine* frequency bands.

Q: J. V. NARLIKAR:

(1) How do you quantify the level of polarization? Is it in terms of intensity ratios (of polarized to unpolarized radiation)?

---

[1] Parenthetically, one also needs the average to be defined, i.e., we need our Universe not to be a certain type of fractal. For that matter the CMB homogeneity, and in particular the weakness of the Sachs–Wolfe effect, $\sim 10^{-4}$ is an extremely good large-scale safeguard.

[2] Data are of course available up to a much larger $z \sim 1$, but at these larger distances, one anticipates seeing the (very weak) effect of expansion, as is indeed the case. The most recent analysis even shows the change from a decelerating to an accelerating phase as expected from dark energy.

[3] Additionally, one should bear in mind that the FIRAS instrument aboard COBE measured the spectrum of the sky with a 7° beam, but the precision did not allow measuring directly the spectrum of the fluctuations. Still, the correlations with the fluctuations first detected by the DMR instrument (aboard the same satellite) were found to be consistent.

(2) Your concluding values of cosmological parameters in terms of assumed scenarios like Lambda CDM, etc., were very precise. Are they disprovable by direct observations? E.g., if tomorrow it gets accepted by astronomers that $H_0 = 55 \pm 5$ km s^{-1} Mpc^{-1}, will it disprove today's most likely scenarios?

A: F. B.:

I *quoted* numbers with errors bars as given by current analyses, but my conclusions were *not* concerning those precise values, which in my view can, and probably will, be somewhat revised. But I did conclude that the cosmogony paradigm was quite robust and consistent, and that the evidence is robust for a model with $\Omega \sim 1/4$ in cold dark matter and $\Omega \sim 3/4$ for something smoothly distributed and slowly evolving (adding up to a total $\Omega \sim 1$, i.e., a close-to-flat spatial geometry), with initial fluctuations mostly adiabatic and close to scale invariant. This is certainly disprovable by observations. And indeed extremely large observational efforts are currently underway to challenge these conclusions, by increasing the accuracy and the control of systematics in experiments that probe different scales in time and space (which also allows challenging general relativity). In particular, supernovae and weak lensing surveys derive constraints from the local Universe, and are fast improving (with ambitious space projects currently in the design for $\sim$2015–2020), while the CMB anisotropies measurements are poised to improve enormously with the results (in $\sim$2010) of the Planck satellite (and a possible successor devoted specifically to polarization in $\sim$2015–2020?). Large Scale Structure surveys (post-SDSS) are also in the planning. We therefore will soon have (again) great opportunities to falsify (or consolidate) the current model by checking whether consistency between different probes remains, when the accuracy is increased by factors of 10, and when new predicted phenomena will be within reach (like the lensing of the CMB by LSS). And maybe we will be lucky enough to find something compelling in the data that will again force us to accept new views on our Universe and the underlying physics.

5

Abundances of light nuclei

Keith A. Olive

William I. Fine Theoretical Physics Institute,
University of Minnesota, Minneapolis, MN 55455, USA

Abstract

An overview of the standard model of big bang nucleosynthesis (BBN) in the post-WMAP era is presented. In this context, the theoretical prediction for the abundances of D, ^{3}He, ^{4}He, and ^{7}Li is discussed. The observational determination of the light nuclides is also discussed. While, the D and ^{4}He observations are concordant with BBN predictions, ^{7}Li remains discrepant with the CMB-preferred baryon density and possible explanations are reviewed.

The standard model [1] of big bang nucleosynthesis (BBN) is based on an extended nuclear network in a homogeneous and isotropic cosmology. Apart from the input nuclear cross sections, the theory contains only a single parameter, namely the baryon-to-photon ratio, $\eta \equiv n_B/n_\gamma$. The theory then allows one to make predictions (with specified uncertainties) of the abundances of the light elements, D, ^{3}He, ^{4}He, and ^{7}Li.

There have been many improvements over the last few years in the state of the theory, particularly in the treatment of the nuclear cross-sections. However, perhaps the most important new input is the WMAP determination of the baryon density [2], $\Omega_B h^2$, or equivalently η. Thus one is now able to make very precise predictions of the light element isotopes, which can be individually compared with observation [3]. The predictions span some nine orders of magnitude in abundance. The major uncertainties in BBN calculations come from the thermonuclear reaction rates. There are 11 key strong rates (as well as the neutron lifetime) that dominate the uncertainty budget [4, 5].

Recently the input nuclear data have been carefully reassessed [4–7], leading to improved precision in the abundance predictions. The NACRE collaboration presented a larger focus nuclear compilation [6]. In an attempt to increase the rigor of the NACRE errors, we reanalyzed [4] the data using NACRE cross-section fits defining a "sample variance," which takes into account systematic differences

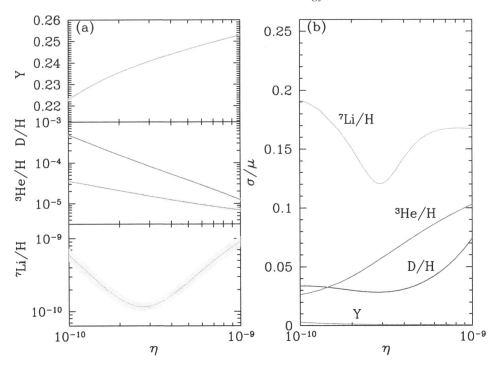

Figure 5.1 The predictions of standard BBN [4], with thermonuclear rates based on the NACRE compilation [6]. (a) Primordial abundances as a function of the baryon-to-photon ratio η. The lines give the mean values, and the surrounding bands give the 1σ uncertainties. (b) The 1σ abundance uncertainties, expressed as a fraction of the mean value μ for each η.

between data sets. For example, notable improvements include a reduction in the uncertainty in the rate for $^3He\,(n,\,p)$T from 10% to 3.5% and for T $(\alpha,\,\gamma)^7$Li from $\sim$23 − 30% to $\sim$4%. Since then, new data and techniques have become available, motivating new compilations.

The resulting elemental abundances predicted by standard BBN are shown in Fig. 5.1 as a function of η [4]. The left plot shows the abundance of ^{4}He by mass, Y, and the abundances of the other three isotopes by number. The curves indicate the central predictions from BBN, while the bands correspond to the uncertainty in the predicted abundances. This theoretical uncertainty is shown explicitly in the right panel as a function of η.

With the increased precision of microwave background anisotropy measurements, it is now possible to use the the CMB to determine independently the baryon density. Allowing for a "running" spectral index lowers the WMAP determination of η. It is [2] $\eta_{10} = 6.14 \pm 0.25$. Equivalently, this can be stated as the allowed range for the baryon mass density today expressed as a fraction of the critical density:

$\Omega_B = \rho_B / \rho_{crit} \simeq \eta_{10} h^{-2} / 274 = (0.0224 \pm 0.0009) h^{-2}$, where $h \equiv H_0 / 100$ km s^{-1} Mpc^{-1} is the present Hubble parameter.

Within the context of the Standard Model (i.e., with $N_\nu = 3$), BBN becomes a zero-parameter theory, and the light element predictions are completely determined to within the uncertainties in η_{CMB} and the BBN theoretical errors. Comparison with light element observations can then be used to restate the test of BBN–CMB consistency, or to turn the problem around and test the astrophysics of post-BBN light element evolution [8].

In recent years, high-resolution spectra have revealed the presence of D in high-redshift, low-metallicity quasar absorption systems (QAS), via its isotope-shifted Lyman-α absorption. It is believed that there are no astrophysical sources of deuterium [9], so any measurement of D/H provides a lower limit to primordial D/H and thus an upper limit on η; for example, the local interstellar value of D/H $= (1.5 \pm 0.1) \times 10^{-5}$ [10] requires that $\eta_{10} \leq 9$. In fact, local interstellar D may have been depleted by a factor of 2 or more due to stellar processing; however, for the high-redshift systems, conventional models of galactic nucleosynthesis (chemical evolution) do not predict significant D/H depletion [11].

The five most precise observations of deuterium [12] in QAS give D/H $= (2.78 \pm 0.29) \times 10^{-5}$, where the error is statistical only. Inspection of the data shown in the figure clearly indicates the need for concern over systematic errors. We thus conservatively bracket the observed values with a range D/H $= 2$–5×10^{-5}, which corresponds to a range in η_{10} of 4–8, which easily brackets the CMB determined value.

Using the WMAP value for the baryon density the primordial D/H abundance is predicted to be [4, 7] $(D/H)_p = 2.55^{+0.21}_{-0.20} \times 10^{-5}$. As one can see, this value is in very good agreement with the observational value.

We observe ^{4}He in clouds of ionized hydrogen (HII regions), the most metal-poor of which are in dwarf galaxies. There is now a large body of data on ^{4}He and CNO in these systems [13]. ^{4}He abundance determinations depend on a number of physical parameters associated with the HII region in addition to the overall intensity of the He emission line. These include the temperature, electron density, optical depth, and degree of underlying absorption.

The question of systematic uncertainties was addressed in some detail in [14]. It was shown that there exist severe degeneracies inherent in the self-consistent method, particularly when the effects of underlying absorption are taken into account.

Recently a careful study of the systematic uncertainties in ^{4}He, particularly the role of underlying absorption, has led to a higher value for the primordial abundance of ^{4}He [15]. Using a subset of the highest quality from the data of Izotov and Thuan [13], all of the physical parameters listed above including the ^{4}He abundance

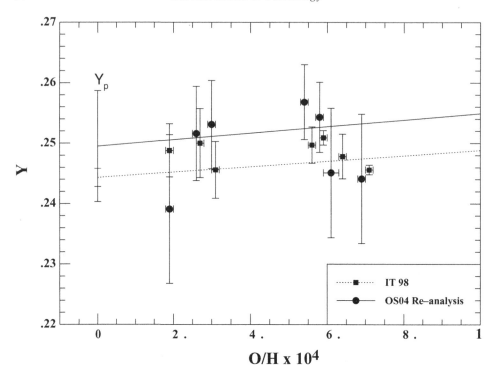

Figure 5.2 A comparison of the results for the best targets [13] and a re-analysis of the spectra for those targets [15].

were determined self-consistently with Monte Carlo methods [14]. Note that the ^{4}He abundances are systematically higher, and the uncertainties are several times larger than quoted in [13]. In fact this study has shown that the determined value of Y_p is highly sensitive to the method of analysis used. The result is shown in Fig. 5.2 together with a comparison of the previous result. The extrapolated ^{4}He abundance was determined to be $Y_p = 0.2495 \pm 0.0092$. The value of η corresponding to this abundance is $\eta_{10} = 6.9^{+11.8}_{-4.0}$ and clearly overlaps with η_{CMB}. Conservatively, it would be difficult at this time to exclude any value of Y_p inside the range $0.232-0.258$.

At the WMAP value for η, the ^{4}He abundance is predicted to be [4, 7] $Y_p = 0.2485 \pm 0.0005$.

The systems best suited for Li observations are metal-poor halo stars in our galaxy. Observations have long shown [16] that Li does not vary significantly in Pop II stars with metallicities $\lesssim 1/30$ of solar – the "Spite plateau." Recent precision data suggest a small but significant correlation between Li and Fe [17], which can be understood as the result of Li production from Galactic cosmic rays [18]. Extrapolating to zero metallicity one arrives at a primordial value [19] $Li/H|_p = (1.23^{+0.34}_{-0.16}) \times 10^{-10}$.

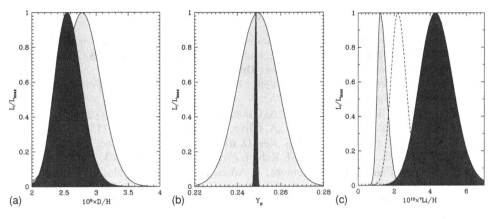

Figure 5.3 Primordial light element abundances as predicted by BBN and WMAP (dark shaded regions) [24]. Different observational assessments of primordial abundances are plotted as follows: (a) the light shaded region shows $D/H = (2.78 \pm 0.29) \times 10^{-5}$; (b) the light shaded region shows $Y_p = 0.249 \pm 0.009$; (c) the light shaded region shows $^7Li/H = 1.23^{+0.34}_{-0.16} \times 10^{-10}$, while the dashed curve shows $^7Li/H = (2.19 \pm 0.28) \times 10^{-10}$.

Recent data [20] with temperatures based on $H\alpha$ lines (considered to give systematically high temperatures) yields $^7Li/H = (2.19 \pm 0.28) \times 10^{-10}$. These results are based on a globular cluster sample (NGC 6397). This result is consistent with previous Li measurements of the same cluster, which gave $^7Li/H = (1.91 \pm 0.44) \times 10^{-10}$ [21] and $^7Li/H = (1.69 \pm 0.27) \times 10^{-10}$ [22]. A related study (also of globular cluster stars) gives $^7Li/H = (2.29 \pm 0.94) \times 10^{-10}$ [23].

The ^7Li abundance based on the WMAP baryon density is predicted to be [4, 7] $^7Li/H = 4.26^{+0.73}_{-0.60} \times 10^{-10}$. This value is in clear contradiction with most estimates of the primordial Li abundance.

In Fig. 5.3, we show the direct comparison between the BBN predicted abundances using the WMAP value of $\eta_{10} = 6.14 \pm 0.25$ with the observations [24]. As one can see, there is very good agreement between theory and observation for both D/H and ^4He. Of course, in the case of ^4He, concordance is almost guaranteed by the large errors associated with the observed abundance. In contrast, as was just noted above, there is a marked discrepancy in the case of ^7Li.

Acknowledgements

I would like to thank R. Cyburt, B. Fields, E. Skillman, and E. Vangioni-Flam for recent (and enjoyable) collaborations on BBN. This work was partially supported by DOE grant DE-FG02–94ER-40823.

References

[1] T. P. Walker, G. Steigman, D. N. Schramm, K. A. Olive, and K. Kang, *Ap.J.*, **376** (1991) 51; K. A. Olive, G. Steigman, and T. P. Walker, *Phys. Rep.*, **333** (2000) 389; B. D. Fields and S. Sarkar, *Phys. Rev.*, **D66** (2002) 010001.

[2] C. L. Bennett *et al.*, *Astrophys. J. Suppl.*, **148** (2003) 1; D. N. Spergel *et al.*, *Astrophys. J. Suppl.*, **148** (2003) 175.

[3] R. H. Cyburt, B. D. Fields, and K. A. Olive, *Phys. Lett. B*, **567** (2003) 227.

[4] R. H. Cyburt, B. D. Fields, and K. A. Olive, *New Astron.*, **6** (1996) 215.

[5] K. M. Nollett and S. Burles, *Phys. Rev. D*, **61** (2000) 123505; A. Coc, E. Vangioni-Flam, M. Cassé, and M. Rabiet, *Phys. Rev. D*, **65** (2002) 043510; A. Coc, E. Vangioni-Flam, P. Descouvemont, A. Adahchour, and C. Angulo, *Ap. J.*, **600** (2004) 544; P. Descouvemont, A. Adahchour, C. Angulo, A. Coc, and E. Vangioni-Flam, *ADNDT* **88**; P. D. Serpico, S. Es-posito, F. Iocco, G. Mangano, G. Miele, and O. Pisanti, *JCAP* 0412 (2004) 010.

[6] C. Angulo *et al.*, *Nucl. Phys. A*, **656** (1999) 3.

[7] R. H. Cyburt, *Phys. Rev. D*, **70** (2004) 023505.

[8] R. H. Cyburt, B. D. Fields, and K. A. Olive, *Astropart. Phys.*, **17** (2002) 87.

[9] R. I. Epstein, J. M. Lattimer, and D. N. Schramm, *Nature*, **263** (1976) 198.

[10] H. W. Moos *et al.*, *Astrophys. J. Suppl.*, **140** (2002) 3.

[11] D. D. Clayton, *Ap. J.*, **290** (1985) 428; B. D. Fields, *Ap.J.*, **456** (1996) 678.

[12] S. Burles and D. Tytler, *Ap.J.*, **499** (1998) 699; *Ap.J.*, **507** (1998) 732; J. M. O'Meara, D. Tytler, D. Kirkman, N. Suzuki, J. X. Prochaska, D. Lubin, and A. M. Wolfe, *Astrophys. J.*, **552** (2001) 718; D. Kirkman, D. Tytler, N. Suzuki, J. M. O'Meara, and D. Lubin, *Ap.J. Supp.*, **149** (2003) 1; M. Pettini and D. V. Bowen, *Astrophys. J.*, **560** (2001) 41.

[13] Y. I. Izotov, T. X. Thuan, and V. A. Lipovetsky, *Ap.J.*, **435** (1994) 647; *Ap.J. S.*, **108** (1997) 1; Y. I. Izotov and T. X. Thuan, *Ap.J.*, **500** (1998) 188.

[14] K. A. Olive and E. D. Skillman, *New Astron.*, **6** (2001) 119.

[15] K. A. Olive and E. D. Skillman, *Astrophys. J.* **617**, 29 (2004)

[16] F. Spite and M. Spite, *A. A.*, **115** (1982) 357; P. Molaro, F. Primas, and P. Bonifacio, *A. A.*, **295** (1995) L47; P. Bonifacio and P. Molaro, *MNRAS*, **285** (1997) 847.

[17] S. G. Ryan, J. E. Norris, and T. C. Beers, *Ap.J.*, **523** (1999) 654.

[18] B. D. Fields and K. A. Olive, *New Astron.*, **4** (1999) 255; E. Vangioni-Flam, M. Cassé, R. Cayrel, J. Audouze, M. Spite, and F. Spite, *New Astron.*, **4** (1999) 245.

[19] S. G. Ryan, T. C. Beers, K. A. Olive, B. D. Fields, and J. E. Norris, *Ap.J. Lett.*, **530** (2000) L57.

[20] P. Bonifacio *et al.*, *Astron. Astrophys.*, **390** (2002) 91.

[21] L. Pasquini and P. Molaro, *A. A.*, **307** (1996) 761.

[22] F. Thevenin *et al.*, *A. A.*, **373** (2001) 905.

[23] P. Bonifacio, *Astron. Astrophys.*, **395** (2002) 515.

[24] R. H. Cyburt, B. D. Fields, K. A. Olive, and E. Skillman, *Astropart. Phys.* **23**, 313 (2005).

Discussion

Q : A. BLANCHARD :

Keith, a few years ago I heard somebody claiming that the slope of the He, He/O correlation was not in good agreement with models of stellar evolution.

A : K.O. :

It has generally been true that the data for He and O/H in low metallicity dwarf galaxies has indicated a value of $\Delta Y/\Delta Z$ that is rather high compared with the value obtained from chemical evolution models. For example, the data of Izotov and Thuan (*Ap.J.*, **602**, 200, 2004) give values of $\Delta Y/\Delta Z$ in the range 3–5 (Z 220 /H). Whereas models tend to give lower values in the range 1–2, models with winds can be constructed to amplify $\Delta Y/\Delta Z$; see, for example, Pilyugin (*A.A.*, **277**, 42, 1993) or Fields and Olive (*Ap.J.*, **506**, 177, 1998).

In the recent analysis of the ^{4}He that I spoke about (Olive and Skillman, astro-ph/0405588), the derived slope is significantly smaller, $\Delta Y/\Delta Z$ 2.4 (with a large uncertainty), and should fit standard models more easily.

Q : J.-C. PECKER :

I just want to draw attention to the fact that abundances are basically determined (especially metallic abundances) from a single-minded theory of stellar abundances, assuming iso-optical depth surfaces to be essentially spherical. We know that this is certainly far from being true for supergiants (which dominate the spectrum of galaxies) leading to errors in the abundances of one or two orders of magnitude.

A : K.O. :

I certainly agree that one of the primary uncertainties in abundances derived from stellar spectra (such as ^{7}Li) lies in the assumed stellar models. As you mention, iron abundances in metal-poor stars often vary by as much as an order of magnitude (for the same star) depending on the observation and assumed parameters used in the stellar model. Perhaps this is the primary cause for the discrepancy in ^{7}Li abundances between the predicted and observationally derived values.

6

Evidence for an accelerating universe or lack of?

Blanchard, A.

LATT, UPS, CNRS, UMR 5572, 14 Av E. Belin 31 00 Toulouse, France

Abstract

There is now a large consensus on the preferred cosmological model, which is known as the concordance model. This model relies on the introduction of a cosmological constant that represents the dominant form of energy densities in the present-day Universe. I briefly discuss the fact that from an astrophysicist point of view the evidence for a cosmological constant, although compelling, is not of sufficient robustness to consider that its existence has been demonstrated beyond reasonable doubt. I present the preliminary results of the Ω project, a large XMM program devoted to observing distant SHARC clusters. For the first time a measurement of the L–T evolution with XMM has been obtained. We found clear evidence for a positive evolution of the L–T relation, in agreement with most previous analysis based on ASCA and Chandra observations. Its cosmological implication is also discussed based on a new analysis of different X-ray surveys: EMSS, RDCS, MACS, SHARC, 160 deg^2. It is found that a high matter density model fits remarkably well all these surveys, in agreement with all existing previous analyses following the same strategy. Concordance models produce far more high redshift massive clusters than observed in all existing X-ray surveys. This failure could indicate a deviation from the expected scaling of the M–T relation with redshift. However, no signature of such a possibility is found in existing data. I conclude that the properties of distant X-ray clusters as evidenced by XMM provide reliable indication in favor of an Einstein–de Sitter universe.

1 Introduction

Two key observations obtained during the last ten years have led to the building of the concordance model: the detection of the Doppler peak, of which first evidence has been obtained by the Saskatoon experiment (Netterfield *et al.*, 1995), which provided the first indication that the Universe was nearly flat (Lineweaver *et al.*, 1997) and the Hubble diagram of distant SNIa, which provides evidence for an accelerating Universe. The observational case for these two results has gained strength over the last decade. Furthermore, the Λ-CDM model was recognized as a satisfying description of the power spectrum of matter distribution over a large scale and consistent with the traditional M/L arguments. Furthermore, it could accommodate the value of the Hubble constant obtained by the HST. The concordance model appeared as a robustly established model. It is fair to recognize that the concordance model is actually in agreement with almost all observations that are relevant to the description of the Universe on a large scale. There is however some room for skepticism: The concordance model requires the introduction of a new component of the Universe, either in the form of a cosmological constant or in the form of quintessence, or some new gravitational physics able to produce a negative gravitational action on large scale and be the dominant contribution to the present day Universe. If true this is certainly one of the most remarkable results of modern cosmology, which is likely to have a deep impact on fundamental physics. However, I would like to emphasize that direct evidence for a non-zero cosmological constant is still weak, and that, contrary to what is generally believed, an Einstein–de Sitter Universe is still a viable option, supported by the properties of distant clusters.

I will not discuss at length the weakness of evidence in favor of the concordance. The reader might have a look at Blanchard (2003) for a discussion on this issue. An updated discussion is also presented in Blanchard *et al.*, (2003).

As I mentioned, the most direct evidence for an accelerating Universe is the Hubble diagram of distant SNIa. Indeed this diagram has been interpreted as providing evidence for an acceleration phase follow by a deceleration phase (Riess *et al.*, 2004). However, if one plots the difference of magnitude (at a given redshift) between a concordance model and an Einstein–de Sitter Universe, it is found that this quantity is nearly linear with time (see Figure 6.1). This illustrates that a (astro)physical term making a time-dependant evolution might mimic an accelerating Universe.

The question of the value of matter density in the present-day Universe is generally considered as a problem essentially solved. However, although the flatness of the Universe has been established beyond reasonable doubt for almost ten years, the only direct evidence for a cosmological constant comes from the Hubble diagram of distant supernovae (and the possible detection of the cross correlation of CMB and surveys of the local matter content: Boughn & Crittenden, 2004). The WMAP signal, for instance, as well as other LSS properties of the Universe can be well reproduced in an Einstein–de Sitter model (Blanchard *et al.*, 2003). I therefore firmly believe that further evidence in favor of the actual existence of a cosmological constant is needed before it can be regarded as an established scientific fact.

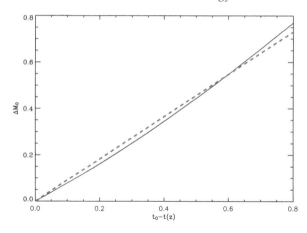

Figure 6.1 Difference of magnitude (at a given redshift) between a concordance model and an Einstein–de Sitter universe versus time (black line). This difference is close to a linear relation (dashed line). The corresponding range in redshift is from $z = 0$ to $z \sim 2$.

The XMM-Ω project (Bartlett *et al.*, 2001) was conducted in order to provide an accurate estimation of the possible evolution of the luminosity–temperature relation at high redshift for clusters of medium luminosity, which constitute the bulk of X-ray selected samples, allowing removal of a major source of degeneracy in the determination of Ω_M from cluster abundance evolution.

2 Observed evolution of the *L–T* relation of X-ray clusters

Lumb *et al.*, (2004), present the results of the X-ray measurements of eight distant clusters with redshifts between 0.45 and 0.62. By comparing them with various local *L–T* relations, clear evidence for evolution in the *L–T* relation has been found. The possible evolution has been modeled in the following way:

$$L_x = L_6(0) \left(\frac{T}{6 \text{keV}} \right)^\alpha (1 + z)^\beta \tag{1}$$

where $L_6(0)(\frac{T}{6\text{keV}})^\alpha$ is the local *L–T* relation. β is found to be of the order of 0.6 in an Einstein–de Sitter cosmology (Lumb *et al.*, 2004; Vauclair *et al.*, 2003). This result is entirely consistent with previous analyses (Sadat *et al.*, 1998; Vikhlinin *et al.*, 2002).

3 Cosmological interpretation

The evolution of the abundance of X-ray clusters is known to be a powerful cosmological test (Oukbir and Blanchard, 1992). Indeed the evolution of the number

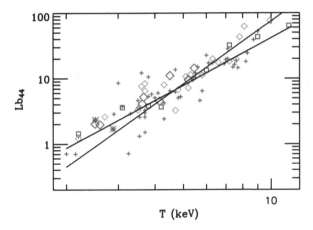

Figure 6.2 Temperature–luminosity of X-ray clusters: Crosses are local clusters from a flux selected sample, gray diamonds are distant clusters from Chandra (Vikhlinin *et al.*, 2002), large dark diamonds are clusters from the XMM Ω project, other dark symbols are other XMM clusters with $z > 0.3$.

of clusters of a given mass is a sensitive function of the cosmological density of the Universe, very weakly depending on other quantities when properly normalized (Blanchard and Bartlett, 1998).

Attempts to apply this test have been performed but still using only a very limited number of clusters (typically 10 at redshift 0.35) (Henry, 1997; Viana and Liddle, 1999; Eke *et al.*, 1998; Blanchard *et al.*, 2000). In Blanchard *et al.*, it was found that $\Omega = 0.86 \pm 0.25(1\sigma)$, so that a concordance model deviates at only a 2-σ level, while systematics differences explain the values obtained from the various authors. On the other hand, number counts allow one to use samples comprising many more clusters. Indeed using simultaneously different existing surveys, EMSS, SHARC, RDCS, MACs NEP, and 160 deg^2, one can use information provided by more than 300 clusters with $z > 0.3$ (not necessarily independent). In order to model the clusters' number counts, for which temperatures are not known, it is necessary to have a good knowledge of the L–T relation over the redshift range that is investigated, information that has been provided by XMM and Chandra. Number counts can then be computed:

$$N(> f_x, z, \Delta z) = \Omega \int_{z-\Delta z}^{z+\Delta z} \frac{\partial N}{\partial z}(L_x > 4\pi D_l^2 f_x) dz$$

$$= \Omega \int_{z-\Delta z}^{z+\Delta z} N(> T(z)) dV(z)$$

$$= \Omega \int_{z-\Delta z}^{z+\Delta z} \int_{M(z)}^{+\infty} N(M, z) dM dV(z) \qquad (2)$$

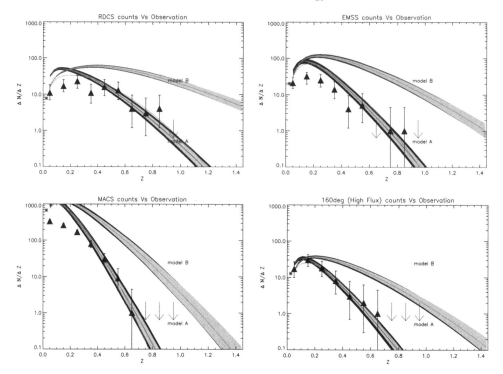

Figure 6.3 Theoretical number counts in terms of redshift ($\Delta z = 0.1$) for the different surveys: RDCS, EMSS, MACS, and 160 deg^2 (high flux, corresponding to fluxes $f_x > 2 \times 10^{-13}$ erg s^{-1} cm^{-2}). Observed numbers are triangles with 95% confidence interval on the density assuming Poissonian statistics (arrows are 95% upper limits). The upper curves are the predictions in the concordance model (model B). The lower curves are for critical universe (model A). Different systematics have been investigated (see Vauclair *et al.*, 2003). Grey areas are uncertainties from uncertainties on σ_8 and on $L-T$ evolution. The – hardly visible!– 3-dotted-dashed lines show the predicted counts in the concordance model using non-standard scaling of $M-T$ relation with redshift.

where $T(z)$ is the temperature threshold corresponding to the flux f_x as given by the observations, being therefore independent of the cosmological model. For most surveys the above formula has to be adapted to the fact that the area varies with the flux limit, and eventually with redshift. Several ingredients are needed: The local abundance of clusters as given by the temperature distribution function ($N(T)$), the mass–temperature relation and its evolution, the mass function, and the knowledge of the dispersion. Uncertainties in these quantities result in uncertainties in the modeling. However, Vauclair *et al.* showed that these sources of systematics uncertainties are comparable to statistical uncertainties, which are

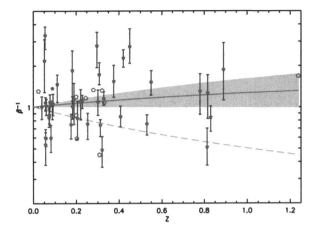

Figure 6.4 The ratio between thermal energy of the gas measured by T_x and the kinetic energy of galaxies measured by their velocity dispersion for a sample of clusters with $T_x \geq 6\,\mathrm{keV}$ with redshift spanning from 0 to 1.2. No sign of evolution is found. The best fit is the continuous line; grey area is the formal one σ region, dashed line is the level necessary to make the concordance in agreement with the X-ray clusters counts.

actually very small. Figure 6.2 illustrates the counts obtained with a standard mass temperature relation: $T = 4\,\mathrm{keV}\,M_{15}^{2/3}(1+z)$, the SMT mass function (Sheth *et al.*, 2001), and the L–T relation observed by XMM with its uncertainty. These counts were computed for different existing surveys to which they can be compared.

4 Possible loopholes

During this analysis we have investigated in great detail numerous possible sources of systematics (local samples, normalization of the M–T relations, local L–T relation, dispersion in the various relations). We have also checked that the local luminosity in our models is in rough agreement with local surveys (without requesting it explicitly). Special attention has been paid to selection function. For instance, if flux limit, or identically flux calibration in faint surveys, is out by a factor of 2–3 the concordance would be much closer to existing surveys. However, typical uncertainty is considered to be of the order of 20%. We have identified only one possible realistic way to reproduce number counts in a concordance model, which is to assume that the redshift evolution of the assumed M–T relation is not standard: $T \propto M_{15}^{2/3}$. This is conceivably possible if a large fraction of the thermal energy of the gas originates from processes other than the gravitational collapse (although it remains to be shown that this is actually possible in a realistic way). It is possible to test observationally this latter possibility. To this aim we have collected some existing measurements of velocity dispersion σ existing for massive clusters (selected to be with temperature greater than 6 keV). The quantity:

$$\beta^{-1} \propto \frac{T_x}{\sigma^2}$$

should evolve with redshift accordingly to $(1 + z)^{-1}$ if the $M-T$ relation evolved accordingly to the above non-standard scheme (and should remain constant in the standard case). We found no sign of the non-standard behavior, which is in principle ruled out near the $3-\sigma$ level.

5 Conclusion

The major results obtained with the Ω project are the first XMM measurement of the evolution of the luminosity–temperature with redshift. A clear positive evolution has been detected, in agreement with previous results including those obtained by Chandra (Vikhlinin *et al.*, 2002). The second important result is that this evolving $L-T$ produces counts in the concordance model that are inconsistent with the observed counts in all existing published surveys. This could be the sign of a high density Universe or a deviation from the expected scaling of the $M-T$ relation with redshift. Our investigation of the ratio T_x/σ^2 does not show any sign of such deviation. The distribution of X-ray selected clusters seems therefore inconsistent with the standard picture of a low cosmological density parameter and rather seriously favors a high density Universe.

References

Bartlett, J. *et al.*, 2001, proceedings of the XXI rencontre de Moriond, astro-ph/0106098
Blanchard, A., 2003, proceedings of the 2002 Jenam Conference, *Ap. & SS.*, **290**, 135.
Blanchard, A. & Bartlett, J., 1998, *A&A*, **332**, 49L
Blanchard, A., Douspis, M., Rowan-Robinson, M., & Sarkar, S., 2003, *A&A*, **412**, 35
Blanchard, A., Sadat, R., Bartlett, J., & Le Dour, M., 2000, *A&A*, **362**, 809
Boughn, S. & Crittenden, R., 2004, *Nature*, **427**, 45
Eke, V. R., Cole, S., Frenk, C., & Henry, P. J., 1998, *MNRAS*, **298**, 1145
Henry, J. P., 1997, *ApJ*, **489**, L1
Lineweaver, C., Barbosa, D., Blanchard, A. & Bartlett, J., 1997, *A&A*, **322**, 365.
Lumb, D. *et al.*, 2004, *A&A*, **420**, 853
Netterfield, C. B., Jarosik, N., Page, L., Wilkinson, D., & Wollack, E., 1995, *ApJ*, **445**, L69
Oukbir, J. & Blanchard, A., 1992, *A&A*, **262**, L21
Riess, A. G. *et al.*, 2004, *ApJ*, **607**, 665

Sadat, R., Blanchard, A., & Oukbir, J., 1998, *A&A*, **329**, 21
Sadat, R. & Blanchard, A., 2001, *A&A*, **371**, 19
Viana, T. P. & Liddle, A. R., 1999, *MNRAS*, **303**, 535
Vauclair, S. C. *et al.*, 2003, *A&AL*, **412**, L37
Vikhlinin, A. *et al.*, 2002, *ApJL*, **578**, 107

Discussion

Q : B. LEMPEL :
What was the density of the Universe when it became transparent?

A : A. B. :
The redshift was 1000 so the density is 1000^3 the present density, i.e., around 10^{-20} g cm^{-3}.

Q : B. LEMPEL :
This seems to me absurd, as matter is transparent in the Sun at the level of the photosphere, i.e., where the density is around 10^{-12} g cm^{-3}.

A : A. B. :
What does matter is the optical thickness, which is the integral of the density along the distance: The Sun becomes opaque over a distance of a few hundred kilometers, it takes several megaparsecs for the Universe to become opaque at $z = 1000$.

Q : K. OLIVE :
How do you account for the value of the baryon fraction in clusters $f_b \sim 0.15$ if $\Omega_m = 1$, given that $\Omega_b \sim 0.045$?

A : A. B. :
In order to fit the WMAP Cl, you need to have $H_o \sim 45\text{--}50$ km s^{-1} Mpc^{-1}. In this case $\Omega_b \sim 0.1$ and $f_b \sim 0.125$. The tension is thus not strong. Sadat and Blanchard (2001) gave a more detailed analysis on this issue.

Q : H. ARP :
In the evolution of clusters, what is evolving? Which properties? Mass, members?

A : A. B. :
The abundance, i.e., the number of clusters of a given temperature per unit covolume, is decreasing rapidly with redshift. From the XMM data we also found a slight increase of the luminosity of clusters for a given temperature.

Q : J.-C. PECKER :

Two elements are involved in the temperature of the microwave radiation: Its intensity I and the wavelength of the maximum of intensity (the Wien temperature). Are they both equal?

In other words, is the MCB radiation that of a good undisturbed black body or of a slightly diluted blackbody?

A : A. B. :

COBE provided the measurements of the intensity of the radiation at 50 different frequencies and all these measurements fall remarkably well on a black body, curve with T = 2.726 K with the Wien region widely measured. Actually no departure from a black body is seen at a level of 10^{-4} (in energy), and the limit of the measurement comes from the on board (reference) black body, which could not be perfectly controlled!

Part III

Standard cosmology

7

Cosmology, an overview of the standard model

Francis Bernardeau

Service de Physique Théorique, CEA/DSM/SPhT,
Unité de recherche associée au CNRS, CEA/Saclay 91191 Gif-sur-Yvette cédex, France

Abstract

The full recognition of the genuine expansion of the Universe and all its consequences is what led to the construction of the "hot big-bang" scenario. The reasons why it is now widely accepted as the standard model of cosmology are reviewed. They are all related to the fact that, if matter and energy are to be conserved in an expanding Universe, its content gets diluted over cosmological time implying that it has experienced a thermal history. Many physical phenomena associated with this evolution have been identified, the signatures of which have been actively looked for and indeed been found in a series of crucial observations, from big-bang nucleosynthesis to the patterns expected from the gravitational growth of structure of the Universe.

What is well established, what is speculative, and the questions that are left totally unanswered in the standard cosmology scenario are succinctly presented.

1 Introduction

Most modern cosmologists would certainly agree that the rapid and recent progress in observational cosmology has put the standard model of cosmology, what is often referred to as the hot-big-bang scenario, on increasingly solid ground. Since the discovery by Hubble in 1929 of the apparent expansion of the Universe from the observed tendency of the faint galaxies to be redshifted, the idea that the Universe is expanding has attracted the attention of many astrophysicists. It led to the idea that the Universe, rather than being in some sense immutable, was born from a primordial explosion, a "Big Bang." As will be discussed in the conclusions, the issue of the "birth" of the Universe is actually beyond the standard cosmology theory. What is left of this idea, however, is that the physical properties of the Universe have rapidly evolved over the course of the cosmological time, leading to a rapid decrease of both the density and the temperature of the Universe.

The discovery of the microwave background by Penzias and Wilson in 1965 can be viewed as the birthdate of modern cosmology. It revealed indeed what is thought of as one of the key relics of the thermal history of the Universe, the photon bath that ought to have emerged when the cosmic plasma temperature dropped below the ionization temperature of neutral hydrogen.

If, at the time of this discovery, the validity of this model could still be largely questioned, nowadays a large number of observations has been collected that puts this somewhat crazy idea – the Universe as a whole has its own history – on firm ground. The aim of this text is to present the central arguments that support this idea. More detailed derivations and presentations of these results can of course be found in cosmology textbooks such as *Gravitation and Cosmology* by Weinberg (on the derivation of basic principles of cosmology from General Relativity [6]); *The Early Universe* by E. Kolb and M. Turner (on the thermal history of the Universe [3]); *Cosmological Inflation and Large-Scale Structure* by Liddle and Lyth [4]; *Modern Cosmology* by S. Dodelson [2], more particularly focused on the physics of inflation and large-scale structure formation.

The second section is devoted to the presentation of the main sequences to be encountered in the thermal history of the Universe, including the big-bang nucleosynthesis and recombination physics, with a special focus on the latter. Recombination time is the time during which the CMB anisotropies were generated and they have become the most precious probe for modern observational cosmology. The third part is devoted to the physics of inflation. The inflationary scenario is not per se part of the standard cosmology model and undoubtedly contains a large proportion of speculative ideas. It nonetheless provides us with a rather convincing mechanism for solving a number of paradoxes that appear in standard cosmology and for the origin of structure. The last section examines the open questions that remain to be solved (and they are numerous) and questions that may be definitely beyond scientific investigations.

2 A brief description of the thermal history of the Universe

2.1 Basic principles and element of cosmography

Before even writing the first equation it is perhaps interesting to remember a few facts. If the Universe is not homogeneous, its metric is nearly locally flat. Except on rare occasions where it can be noticed, photons are basically traveling along straight trajectories. Only rarely can we see effects of gravitational lensing that betray the existence of rather modest (most of the time at 10^{-5} level) metric fluctuations. It is therefore legitimate to speak of the global metric properties of the observable Universe.

If in standard cosmology the global properties of the Universe are evolving with time, the Universe is nonetheless assumed to be, at least statistically, isotropic and homogeneous. In the framework of General Relativity (GR) it is implied that its metric should be of the family of the Friedmann–Robertson–Walker metric,

$$ds^2 = c^2\,dt^2 - a^2(t)\frac{dx^2}{1-kx^2}. \tag{1}$$

It allows the Universe to expand at a speed that depends closely on the energy density of the Universe. The immediate observational consequence is that distant objects appear to be redshifted by an amount proportional to the ratio of the expansion factor a between emission and reception time. Indeed they do. Does it mean the Universe is really evolving with time? Matter and energy conservation then imply that the density of the Universe should be decreasing as,

$$\rho_{mat.} \propto a^{-3} \quad \text{and} \quad \rho_{rad.} \propto a^{-4} \tag{2}$$

respectively. Such an evolution of the physical properties of the Universe is bound to have many consequences. Standard cosmology is to a large extent the confrontation of those predictions with observations.

As the temperature of the Universe drops, different phase transitions can take place. This is summarized in Fig. 7.1, which shows the main events that are thought to have taken place during the thermal history of the Universe. The larger the temperature or the density is, the more speculative these events are. In particular we have no known mechanisms for the baryogenesis, inflation is a very speculative era . . . On the other hand, big-bang nucleosynthesis and the physics of recombination are well-understood, low-energy physics effects. They provide us with the best observational supports for standard cosmology.

2.2 *Big-bang nucleosynthesis*

The big-bang nucleosynthesis (BBN) describes the freezing of the nuclear reaction as the temperature of the cosmic fluid decreases, allowing the creation of heavy nuclei by proton or neutron capture from a plasma made only of electrons, protons, and neutrons (see Olive, these proceedings, Chapter 5, for a more detailed account of these processes). It is to be noted that these reactions are actually out of equilibrium. Then the end result of the freezing process depends intimately on the time rate of this evolution, that is, on the actual value of the Hubble constant H. That BBN can explain the mass fraction of helium that can be measured with a reasonable accuracy, is therefore a non-trivial result.

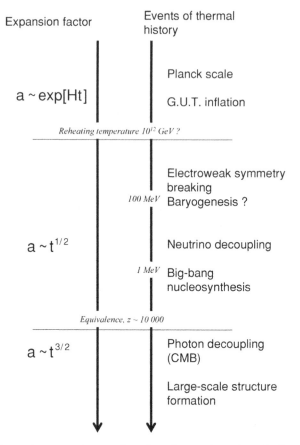

Figure 7.1 The main events that are thought to have taken place during the thermal history of the Universe (right) and behavior of the expansion factor during those stages (left).

The case for BBN has been made even stronger. Indeed it can only work for a small range of number density of nuclei per photon. Until recently this baryon density could only be inferred from rough accuracy measurements (in galaxy or cluster surveys). It has now been determined with a much better accuracy from Cosmic Microwave Background observations (see next section). The agreement between the two approaches is most reassuring for the standard cosmology model supporters!

2.3 Recombination physics

Recombination is the stage during which the plasma temperature drops below the ionization temperature of the nuclei. As the number of free electrons dramatically drops, the photons can freely propagate. The relic of these photons is the Cosmic

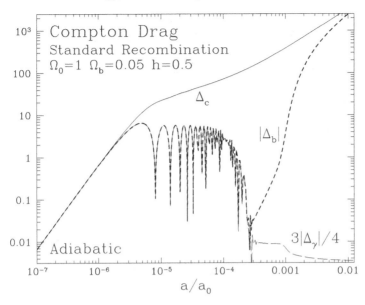

Figure 7.2 Compared evolution of the intrinsic temperature, Δ_γ, of dark matter density, Δ_c, and baryon density, Δ_b, for a proper gauge choice. The corresponding wavelength is at very small scale and the picture exhibits successively the joint adiabatic evolution, the plasma oscillations for the baryon–photon fluid, the Silk damping effects and finally, after decoupling, the fall of the baryons into the dark matter potential wells (figure from [4]).

Microwave Background (CMB). One expects these photons to be isotropically distributed and to exhibit a near perfect black-body spectrum.

These features have indeed been observed. They suggest that a plasma at thermal equilibrium once existed. Still better, the recent observations of the CMB anisotropies have provided us with detailed measurements of the plasma properties such as its sound speed.

The CMB anisotropies actually reveal the plasma oscillations as they reach the last scattering surface. The main equation that drives the energy density modes of the plasma oscillations essentially reads

$$\ddot{\Delta}_k + 2H\dot{\Delta}_k = \frac{3}{2}H^2\Delta_k - c_s^2 k^2 \Delta_k \tag{3}$$

the solution of which obviously depends on the sound speed of the plasma fluid, $c_s = \sqrt{\delta p/\delta\rho}$. The core of CMB anisotropy calculations is then the resolution of the growth of perturbation modes as the content of the Universe (and therefore c_s) evolves. Note that such calculations are linear order calculations in an expansion with respect to the metric fluctuations. They are thus a priori 10^{-5} precision calculations. One example of such a mode growth is depicted on Fig. 7.2 corresponding to a rather small scale mode.

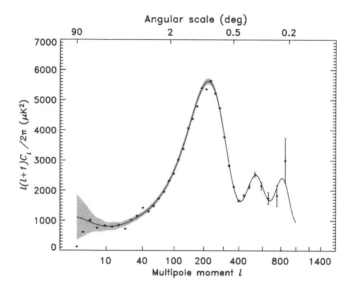

Figure 7.3 The angular spectrum of the temperature anisotropies as measured by WMAP.

When transcribed into the CMB anisotropies those density fluctuations give birth to the CMB anisotropies. The latter are actually the superposition of different effects: The intrinsic temperature fluctuations on the last scattering surface, the metric fluctuations that induce gravitational Doppler effects, and the proper motion of the plasma along the line of sight responsible for kinetic Doppler effects. The resulting C_l (amplitude of modes in an harmonic decomposition, see Bouchet, Chapter 4, these proceedings, for more details) is shown as a solid line on Fig. 7.3 together with the observations provided by the WMAP satellite.

Those results clearly exhibit the expected oscillatory features. They signal among other things the value of the plasma sound speed at recombination time, and incidentally the fraction of baryons it contained at that time.

2.4 The emergence of the concordance model

These observations (together with many others but probably set on less firm grounds) led to the elaboration of a concordance model that reconciles a number of very different observations and constraints, from supernovae surveys, CMB anisotropies, large-scale structure of the Universe, age of oldest stars, etc. The basic parameters of this concordance model are

- $H_0 = 71 \pm 4 \, \text{km s}^{-1} \, \text{Mpc}^{-1}$;
- $\Omega_m h^2 = 0.135 + 0.008 - 0.009$;

- $\Omega_b h^2 = 0.0224 \pm 0.0009$;
- $\Omega_{tot.} = 1.02 \pm 0.02$;
- $n_s = 0.93 \pm 0.03$ at wavenumber $k_0 = 0.05$ Mpc^{-1}

Probably the most striking aspect of this model is that it provides us with a coherent picture for the large-scale structure of the Universe, from CMB fluctuations measured on the last scattering surface to low redshift galaxy catalogs.

As it appears from the above numbers, standard cosmology clearly calls for a dark matter component. This is essential for explaining the shape of the observed matter power spectrum (the reason for that can be found in Fig. 7.2: baryons would not fall into the dark matter potentials if they did not exist!). The case for a cosmological constant ($\Omega_{tot.} > \Omega_m$) is less strong and depends to some extent on assumptions on the initial metric fluctuation index (see Blanchard, Chapter 6, these proceedings) when inferred from CMB observations or to weakly known stellar physics when inferred from supernovae observations.

2.5 The standard cosmological model on a test bench

Skepticism should be a rule in science. Can we further test the standard cosmological model? One important ingredient of standard cosmology is that the large-scale structure of the Universe is here assumed to emerge from gravitational instabilities, e.g., from the amplification of the primordial metric fluctuations into gravitationally bound objects. The CMB–LSS confrontation presented in Fig. 7.4 is basically a test of the linear growth rate of the density fluctuations. As the density contrasts grow, other phenomenological effects can take place. Mode couplings effects in particular induce non-Gaussian properties of the density field that can be explicitly measured (see [1] for details).

The gravitational instability picture in particular implies that, in the case of Gaussian initial perturbations, the leading order term of the three-point correlation function takes the form,

$$\xi_3(\mathbf{x}_1, \mathbf{x}_2, \mathbf{x}_3) = \left[\frac{10}{7} \xi(x_{13})\xi(x_{23}) + \nabla\xi(x_{13}) \cdot \nabla^{-1}\xi(x_{23}) \right.$$
$$+ \nabla\xi(x_{23}) \cdot \nabla^{-1}\xi(x_{13})$$
$$\left. + \frac{4}{7} \left(\nabla_a \nabla_b^{-1}\xi(x_{13})\right)\left(\nabla_a \nabla_b^{-1}\xi(x_{23})\right) \right] + \text{cyc.} \quad (4)$$

when expressed in terms of the two-point correlation function. And indeed in Fig. 7.5, measured three-point function is successfully compared with theoretical predictions. It enforces the idea that the large-scale structure indeed emerged from gravitational instability. Further tests along this direction will be possible in the

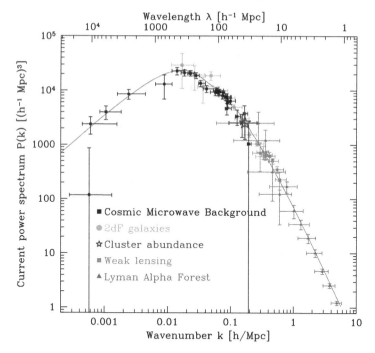

Figure 7.4 The density power spectra as measured from different observations – galaxy catalogs, number density of clusters, cosmic shear measurements and Lyman-α clouds – compared with what should be expected from CMB anisotropies in a Λ–CDM model.

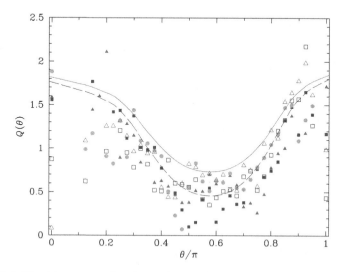

Figure 7.5 The bispectrum for the PSCz catalog for triangles with $0.2 \leq k_1 \leq 0.4$ h Mpc^{-1} and with two sides of ratio $k_2/k_1 = 0.4-0.6$ separated by angle θ. Taken from [3].

near future with the advent of large cosmic shear surveys that directly map the large-scale dark matter distribution.

3 The speculative parts

3.1 The energy content of the Universe

If the standard cosmology picture is to be correct, the observations demand a large fraction of matter as a form of dark, essentially non-interacting matter.

Theorists are not short of ideas for the nature of this dark matter, from Weakly Interactive Massive Particles (WIMPS), Axions, Kaluza–Klein particles . . . However, none of these particles have proved actually to exist. Searches are in progress.

The nature of dark energy is more elusive. If it were to be confirmed it might demand a more radical change of our picture. For instance, we do not know if it corresponds to the energy associated with a scalar field (similar to the inflaton field) or if it corresponds, for instance, to a modification of gravity at large scale.

3.2 The inflationary Universe

Standard cosmology also does not provide us with a mechanism for the generation of the primordial adiabatic fluctuations. Inflation, as it turns out, is the only known paradigm that can give satisfactory answers.

Basically, inflation is an acceleration phase that the Universe might have experienced in the past. Such a phase would be very useful for a number of reasons

- We do not see any monopoles that ought nonetheless to exist in Grand Unified Theory of particle physics;
- The observed temperature of the last scattering surface is isotropic despite the fact that, above 1°, those different patches of the sky would be causally disconnected. This is the Horizon problem;
- The Universe is observed to be nearly flat. A feature that requires extremely fine-tuned parameters at earlier times.

How was such an acceleration stage produced? A solution exists that demands that GR theory and Quantum Field Theory (QFT), the principles of which have both been tested to a high accuracy, be somehow married together. In effect it amounts to exploring the consequences of the dynamics of a simple scalar field in a curved background space-time from the action,

$$S = \int d^4x \sqrt{-g} \left(\mathcal{L} - \frac{\mathcal{R} \, m_{\text{pl.}}^2}{16 \pi} \right) \tag{5}$$

where $\mathcal{L}$ is the Lagrangian density associated with a minimally coupled scalar field,

$$\mathcal{L} = \frac{1}{2} \left(\partial_\mu \varphi \, \partial^\mu \varphi \right) - V(\varphi) \tag{6}$$

The motion equation for φ is naturally obtained from the variational principle, which gives

$$\ddot{\varphi} + 3 \frac{\dot{a}}{a} \dot{\varphi} - \frac{1}{a^2} \Delta\varphi = -\frac{dV}{d\varphi} \tag{7}$$

and the motion equation for the background, that is, for the expansion factor, is similarly obtained from a variational principle. It leads to

$$\dot{a}^2 = \frac{8\pi \, a^2}{3 \, m_{\text{pl.}}^2} \left[\frac{\dot{\varphi}^2}{2} + \frac{(\nabla\varphi)^2}{2} + V(\varphi) \right] \tag{8}$$

In the context of cosmology it is fair to assume, in connection to one of the early remarks in Section 2, the near flatness of the local metric, that the field φ can be decomposed into two parts, its spatial expectation value and its fluctuating part:

$$\varphi(t, \mathbf{x}) = \varphi_0(t) + \delta\varphi(t, \mathbf{x}) \tag{9}$$

Then one assumes that GR can be classically applied to the dynamics of the field expectation value, whereas its fluctuating part will be treated with a quantum field theory point of view, e.g., applying the second quantification rules to $\delta\varphi(t, \mathbf{x})$. Needless to say there is no rigorous mathematical justification for doing so. Quantum gravity does not yet exist as a theory. This approach, however, can be argued to be rather reasonable. The metric fluctuations are indeed known to be small (10^{-5} level) suggesting that the mechanisms that have given birth to the primordial metric fluctuations took place after the Planck era, outside the genuine quantum gravity regime.

The motion equation for φ_0 eventually gives,

$$\ddot{\varphi}_0 + 3 \frac{\dot{a}}{a} \dot{\varphi}_0 = -\frac{dV}{d\varphi}(\varphi_0) \tag{10}$$

A satisfactory inflationary phase will take place if the potential of the energy density is sufficiently large with respect to the kinetic part. This will be possible only if the potential is flat enough. This is quantitatively described in the Slow Roll parameters, $\epsilon = \frac{m_{\text{pl.}}^2}{16\pi} \left(V'/V \right)^2$ and $\eta = m_{\text{pl.}}^2 \, V''/(8\pi \, V)$. Actually it is possible to build a model of inflation arbitrarily close to a de Sitter phase, e.g., constant energy density, provided that the so-called Slow Roll parameters are small enough. Unlike normal matter (whose pressure is always positive) scalar fields can have an effective negative pressure, therefore providing us with one possible inflationary mechanism. The physical interpretation is that the quantum state of the Universe does not correspond to localized φ particles but to a large superhorizon Bose condensate. That such things do exist is certainly speculative but still possible in the realm of QFT.

3.3 The origin of structure

One attractive aspect of inflation is that it provides, with the very same ingredients, a mechanism for the origin of the large-scale structure of the Universe. Basically, in an inflationary Universe, quantum fluctuations of a scalar field can be squeezed and frozen to give birth to a classical stochastic field. The computation of this effect is actually relatively straightforward in the context of quantum field theory. As usual, field fluctuations can be decomposed into mode operators,

$$\delta\varphi = \int d^3 \mathbf{k} \left[a_\mathbf{k} \, \psi_\mathbf{k}(t) \, \exp(i\mathbf{k}.\mathbf{x}) + a_\mathbf{k}^\dagger \, \psi_\mathbf{k}^*(t) \, \exp(-i\mathbf{k}.\mathbf{x}) \right] \tag{11}$$

where $a_\mathbf{k}^\dagger$ and $a_\mathbf{k}$ are the creation and annihilation operators of the field particles of momentum $\mathbf{k}$. They are assumed to obey the commutation rule,

$$[a_\mathbf{k}, a_{-\mathbf{k}'}^\dagger] = \delta(\mathbf{k} + \mathbf{k}') \tag{12}$$

The time dependent mode coefficients $\psi_\mathbf{k}$ are solutions of the motion equation,

$$\ddot{\psi}_\mathbf{k} + 3\,H\,\dot{\delta\varphi} + \frac{k^2}{a^2}\psi_\mathbf{k} = -V''\psi_\mathbf{k} \tag{13}$$

Note that the r.h.s. of this equation vanishes in the Slow Roll approximation (small value of η). The motion equation then corresponds to that of a free massless scalar field. What is important to keep in mind is that the mode amplitudes are determined solely by quantum physics. They are such that they reproduce the expected local commutators in the small-scale Minkowski limit. It leads to a well determined expression of the mode amplitude at large scale (proportional to $H/k^{3/2}$) and consequently of the scalar metric fluctuations: inflaton fluctuations induce superhorizon time laps that are nothing but the scalar metric fluctuations. It is, by the way, worth noting that the resulting metric fluctuations are even larger, that φ_0 is slowly varying (time laps are inversely proportional to $V'(\varphi_0)$), and that the resulting metric amplitudes are at first view scale invariant (same amplitude for all scales) although not exactly, since the value of H is – slowly – evolving as the modes cross the horizon.

At the same time any other degree of freedom of light fields can also give birth to superhorizon quantum fluctuations. This is the case in particular for the tensor modes, the gravitational waves, the amplitudes of which are therefore related to the scalar perturbation.

Consequently generic inflationary models contain the required ingredient to pro-duce primordial metric fluctuations. A number of properties are then expected.

- The metric fluctuations are adiabatic (identical in all cosmic fluid components) and obey Gaussian statistics.
- They have an almost (but not exactly) scale invariant power spectrum.
- One also expects tensor fluctuations, the amplitudes of which trace the energy scale of inflation.

The first two items of this list have indeed been verified in current data sets. The third is next in line but requires much better detector sensitivity. It is also to be noted that the detection of a departure from scale invariant spectrum or gravity waves would strongly support inflation but no detection of such effects would be inconclusive.

Nowadays inflation is the only working paradigm providing an explanatory scheme for the origin of structure. The nature of the inflaton field however still evades theoretical investigations. A number of scenarios have been proposed. They include generic chaotic inflation, F-term or D-term hybrid inflation from SUSY models, superstring tachyonic inflation, etc . . .

4 Conclusions

So what is the status of standard cosmology? A number of ingredients of it have been put on solid ground by recent observations.

There is little doubt for instance that a baryon–photon plasma once existed. Its relic can be found in the CMB whose angular and spectral properties are precisely those that were expected.

The growth of structure also proved to follow patterns expected from the gravi-tational instability picture. That can be verified at many levels, from the CMB–LSS correlations to the mode couplings patterns that have been found to match the expected properties.

Inflation, on the other hand, is still speculative. Only its most generic features have been explicitly verified and we are still waiting for the verification of more specific predictions. What is dramatically missing here is the identification of the inflaton field.

Finally there are questions that standard cosmology hardly addresses and for which there are certainly no clear answers. For instance, standard cosmology does not claim that there actually existed a genuine big bang, i.e., a space-time singularity. For instance in "pre-big-bang" or ekpyrotic models there is no such global GR singularity. In essence, cosmology is essentially the theory of a fluid expansion, it does not say much about the initial impetus that might be at its origin. Whether the Universe is finite or infinite is also a question that is mainly left unanswered. There might exist compact spatial directions at scale larger than the observable Universe that we may never be able to detect. The global space-time structure of the Universe, at scale much beyond the observable Universe, is also unknowable. In the realm of quantum gravity the answer to this question might be extremely complicated!

References

[1] Bernardeau, F., Colombi, S., Gaztañaga, E., Scoccimarro, R., 2002, *Phys. Rept.*, **367**, 1–248
[2] Dodelson, S., *Modern Cosmology*, Academic Press, 2003
[3] Kolb, E. W., Turner, M. S., *"The Early Universe"*, Frontiers in Physics, Reading, MA: Addison-Wesley, 1988
[4] Liddle, A. R., Lyth, D. H., *Cosmological Inflation and Large-Scale Structure*, Cambridge University Press, 2000
[5] Mellier, Y., *Annu. Rev. Astron. Astr.*, **37**, 1271891999
[6] Weinberg, S., *Gravitation and Cosmology: Principles and Applications of the General Theory of Relativity*, New York: Wiley, 1972

Discussion

Q: J. NARLIKAR:

If there is no big bang how did the universe acquire such high energy as at inflation?

A: F. B:

This question is partially addressed in the chaotic inflation picture: quantum fluctuation is all we need. The full answer to this question is, however, well hidden in the mysteries of quantum gravity. But once again standard cosmology does not give much clue on the existence of a genuine "Big Bang."

Q: M. MOLES:

The redshift explanation in the standard model derived from the metric made the assumption of homogeneity and isotropy. The Hubble law is found already at scales that are much smaller than the homogeneity scales. How is that possible?

A: F. B:

First of all the H fluctuations, in linear theory, identify with the divergence field and are proportional to the density fluctuations. The amplitude of H fluctuations therefore diminishes then much more rapidly than those of the velocity. Having said that, that the measured expansion in our close vicinity gets rapidly close to the Hubble value might be a mere coincidence.

Q: H. BROBERG:

The suggestion is that the most straightforward model would be an information front of the Universe that expanded with velocity c from the moment of random creation of an elementary particle and time. Inside the front the Schwarzschild geometry is automatically generated from which G and other parameters are developed. It is explained in a paper published in a recent book on Mach's principle and the origin of inertia, "*Mass and Gravitation in a Machian universe*". The model ensures a "flat" universe and no need of inflation.

A: F. B.: OK.

8

What are the building blocks of our Universe?

Kameshwar C. Wali

Physics Department, Syracuse University
Syracuse, NY 13244-1130, USA

Abstract

We are told that we are living in a Golden Age of Astronomy. Cosmological parameters are found with unprecedented accuracy. Yet, the known form of matter forms only a small fraction of the total energy density of the Universe. Also, a mysterious dark energy dominates the Universe and causes acceleration in the rate of expansion.

1 Introductory remarks

We live in an exciting age of astronomy. Some thirty years ago, cosmology was a science of only two parameters, the current expansion rate or the Hubble constant, H_0, and its change over time or the deceleration parameter, q_0. Questions such as the age of the Universe, its large- and small-scale structure, origin of galaxies, and the formation of stars were considered as speculative with no direct connection to precise measurements. The situation has changed drastically with the discoveries of giant walls of galaxies, voids, dark matter on the one hand, and on the other hand, the tiny variations in the cosmic background radiation and a "mysterious" uniformly distributed, diffuse dark energy causing acceleration of the expansion rate of the Universe. There are some sixteen cosmological parameters whose measured values exhibit unprecedented accuracy in the history of astronomy. Ten of these parameters are "global" in the sense that they pertain to the idealized standard model of a homogeneous isotropic universe governed by the Friedmann–Lemetre–Walker–Robertson metric within the framework of general relativity. The other six refer to more details of the model, to the deviations from homogeneity and their manifestations in the cosmic structure. These numbers are tied to a fundamental theory – big bang, inflationary theory – and it is believed by the practitioners that it accounts for *the origin of structure and geometry of the universe, as well as describing its evolution from a fraction of a second*.

101

In the words of Freedman and Turner [1], the still evolving and emerging picture is described as follows:

In a tiny fraction of a second during the early history of the Universe, there was an enormous explosion called inflation. This expansion smoothed out wrinkles and curvature in the fabric of space-time, and stretched quantum fluctuations on subatomic scales to astrophysical scales. Following inflation was a phase when the Universe was a hot thermal mixture of elementary particles, out of which arose **all the forms of matter that exist today**. *Some 10 000 years into its evolution, gravity began to grow the tiny lumpiness in the matter distribution arising from quantum fluctuations into the rich cosmic structures seen today, from individual galaxies to the great clusters of galaxies and superclusters.*

However, there are wrinkles and surprises in this rosy theoretical picture! [2]. Most of the Universe is made of something fundamentally different from the ordinary matter that we know of. Some 30% of the total-mass energy density is dark matter, whose nature we do not know, but in all likelihood, it is composed of particles formed in the early Universe. About 66% is in the form of a smooth, uniformly diffused energy called the dark energy, whose nature we do not know, but we conjecture that its gravitational effects are responsible for the recently observed acceleration in the rate of expansion of the universe [2]. Approximately only 4% is composed of ordinary matter, the bulk of which is dark. Finally, cosmic microwave background radiation contributes only 0.01% of the total, but it encodes ***information about the space-time structure of the Universe, its early history, and probably even about its ultimate fate***.

In light of this, one wonders whether our present fundamental theories of elementary particles that are supposed to be the building blocks of the Universe are of any relevance to the emerging picture of the Universe. In this review, I present certain aspects concerning the current status of particle theory and its link to cosmology.

2 Beyond the Standard Model; Grand Unified Theories

The current theory of fundamental interactions is the so-called Standard Model, a non-Abelian Yang–Mills type theory based on the gauge group $S(U(3) \times U(2))$ with spontaneous symmetry breaking, induced by a fundamental scalar, called the Higgs meson. It presents a unified theory of weak and electromagnetic interactions (electro/weak) marked by spontaneous symmetry breaking. Strong interactions are described by the gauge theory based on the group $SU(3)$ (Quantum Chromodynamics). It has been enormously successful in its confrontation with experiments. Yet, it is far from a fundamental theory for a number of reasons. First and foremost, it has a large number of free parameters. The starting point is three families of quarks and leptons with their masses totally arbitrary ranging over several orders of magnitude. The theory is renormalizable, but it has quadratic divergences requiring "fine tuning" of the parameters in successive orders of perturbation. It can accommodate CP violation, but has no natural explanation for its origin or the order of magnitude of its violation.

Nonetheless, its enormous success led to its natural extension seeking unification of all the three fundamental interactions, weak, electromagnetic, and strong: *Grand Unified Theories (GUTS)*. In its most pristine form, a Grand Unified Theory postulates that the description of interactions among elementary particles will simplify enormously at some very high energy $E > M_G$ (Grand Unification Mass). The electro/weak and strong interactions, which are the basic interactions at low or present laboratory energies, will be seen as different aspects of one basic interaction among *a set of basic constituents of all matter.* Correspondingly, as one moves up in energy, a symmetry larger than the standard model gauge group $S(U(3) \times U(2))$ will progressively unfold itself, becoming fully manifest at energies exceeding M_G. Initial analysis based on renormalization group methods suggested strongly that the coupling constants that change as a function of energy (a feature of non-Abelian gauge theories) evolve to a unification point at energies around 10^{15} GeV. Since any such unification demanded quarks and leptons to be treated on the same footing, quark–lepton transitions at such energies and above became theoretically mandatory, leading to the possible violation of the well established baryon- and lepton-number conservation laws at low energies. A dramatic consequence was the possibility of observing proton decay! The simplest extension of the Standard Model based on the gauge group $SU(5)$ predicted a lifetime of 10^{29} years for the proton and led to a number of experiments that failed to detect it and have set a limit to proton lifetime beyond 10^{32} years. More complicated models based on bigger simple groups ($SO(10)$, for instance), semi-simple products of groups, and exceptional groups (such as E_6) were proposed and were partially successful in extending the predicted lifetime of the proton and predicting new exotic species of particles.

However, to obtain a full display of the new interactions and to put them to experimental test, we need energies of the order of 10^{15} GeV and greater, which are clearly beyond the present or future terrestrial accelerators. It became evident that astrophysics and cosmology were the natural arena for testing these ideas. In the current popular standard model cosmology, based on the Friedmann–Lemetre–Walker–Robertson metric, the early Universe was in a hot dense phase with temperatures exceeding 10^{16} GeV in its first 10^{-35} seconds after the big bang. The Universe in its early stages was like a giant accelerator and one expected a copious production of all the particles we know, and those we do not know – the super heavy particles predicted by Grand Unified Theories. One could then trace the effects of the new particles and their interactions through the subsequent adiabatic cooling of the Universe down to the present epoch and compare them with astrophysical measurements. It was the beginning of a symbiotic relation between particle physics and astrophysics.

3 Beyond the Standard Model; supersymmetry

Supersymmetry goes beyond the conventional distinction between fermions (odd integral multiples of spin 1/2 particles) as fundamental constituents of matter and bosons (integral multiples of spin 1 particles) as carriers of interactions. It treats both on an equal footing, combining them in a supermultiplet that allows symmetry transformations between them. Conventional space-time symmetries are supplemented by anti-commuting operators that transform fermion into boson and vice versa. Thus it may be looked upon as unification of matter and interactions.

Its main points are as follows.

- Each chiral fermion (quark, lepton) in the Standard Model is accompanied by a spin zero boson (squark, slepton). Likewise each gauge boson and Higgs scalar is accompanied by a spin 1/2 fermion (gaugino, Higgsino).
- All superpartners of Standard Model (SM) particles are *new particles*
- No known SM particle is a superpartner of another SM particle. If supersymmetry were exact, a particle and its superpartner that have the same quantum number should be degenerate in mass.
- Supersymmetry is an *approximate* symmetry of nature. If it were exact, superpartners of SM particles would have been discovered along with the SM particles since they would have been degenerate in mass.

From a theoretical point of view, supersymmetry is very appealing. It is a beautiful symmetry, but it is approximate. There is no unique or elegant symmetry-breaking mechanism. In principle, it has the potential of solving some theoretical problems associated with the quadratic divergences and fine tuning problems generic to the Standard Model and Grand Unified Theories, which invoke spontaneous symmetry breaking through fundamental scalar particles. There is enough freedom in models to meet the experimental limits on proton lifetime exceeding 10^{32} years. The so-called Minimal Supersymmetric Standard Model (MSSM), an extension of the Standard Model, provides more convincing evidence for the unification of all interactions (excluding gravity) than Grand Unified Theories alone. From the point of view of cosmology and astrophysics, broken supersymmetry offers a candidate for dark matter, the "neutralino." [3]

4 Nature of dark matter; candidates for dark matter

Observationally, dark matter appears to be distributed diffusively in external halos around individual galaxies or in a sea through which galaxies move. Here are some speculations concerning its nature.

- It is believed to consist of hypothetical particles called WIMPS (Weakly Interacting Massive Particles), produced probably in the early Universe.
- Their masses should be around electro/weak symmetry-breaking scale, in the 10 GeV–1 TeV range. They should have *neither strong nor electromagnetic* interactions with the known SM particles. If they did, the argument goes, they would have dissipated energy and relaxed to more concentrated structures, where only known baryons are found.
- They must be **Cold**, in the sense that they move slowly with non-relativistic velocities, as opposed to **hot**, light particles moving with relativistic velocities. *Hot* and *Cold* dark matter lead to different predictions regarding galaxy formation. Galaxies are formed first owing to cold dark matter before forming superclusters, whereas the opposite happens with hot dark matter.

It is remarkable that from the simple starting point of cold dark matter and inflation-induced lumpiness, one can envisage a highly successful picture of formation of structure in the universe. From the point of view of particle physics, there are three possible candidates for dark matter:

- **Neutrinos**: The idea that neutrinos could be candidates for dark matter has been there for a long time. They certainly exist in large numbers (roughly one billion for every photon) and they could contribute a huge mass to the dark matter if they were massive enough. Recent experiments on solar and atmospheric neutrino oscillations have established that one or more than one of the types of neutrinos must have a mass. However, neutrino oscillation experiments probe only the mass differences. There are many theoretical models and many experiments to determine their absolute masses. Cosmological observations will play a very important role in setting the absolute scale of neutrino mass just as primordial nucleosynthesis set a limit on the number of light neutrinos. This is because, as mentioned above, hot and cold dark matter predict entirely different courses for the evolution of the large-scale structure. If all the neutrinos are light with masses of an electron volt or less, they constitute hot dark matter. Then, there is a stringent limit on the amount of hot dark matter in order that it does not wipe away the required small-scale structure.
- **Axions**: The axion is probably the first candidate for dark matter that was proposed. A search for it has been going on for quite some time. It has its origin in the theoretical solution of CP violation in strong interactions due to the complex nature of the vacuum in the theory of strong interactions based on quantum chromodynamics (QCD). A global-axial symmetry known as Pecci–Quinn symmetry solved the problem, but it made it necessary to have a massive particle with strong interactions with ordinary matter. When experiments failed to detect such a particle, a mechanism proposed by Dine, Fisher, and

Schrednicki allowed the coupling to matter as well as its mass to be arbitrarily small. The *axion exists, but it cannot be seen.*

Two different mechanisms have been proposed for their production in the early Universe: (a) At the QCD phase transition, when free quarks get bound to form hadrons, a Bose condensate of axions form and these very cold particles behave as cold dark matter. (b) Decay of cosmic strings at the Pecci–Quinn phase transition can also give rise to axions.

Axions are potentially detectable through their weak couplings to electromagnetism. In the presence of a strong magnetic field, the axionic dark matter can decay into two photons. Several new experiments based on cryogenically cooled cavities and the use of an atomic beam of Rydberg atoms as a detector are in progress.

• **Neutralinos**: Broken supersymmetry combined with the conservation of what is called R-parity provides an ideal candidate for dark matter. The lightest particle is absolutely stable and has the necessary properties to form dark matter. In MSSM, the spin-1/2 neutral gauge eigenstates mix and form mass eigenstates after symmetry breaking. These are called *neutralinos*. The lightest among these is considered to be the most probable candidate for dark matter.

Neutralinos are Majorana particles. Their mass estimates in MSSM depend upon five parameters. In order to estimate their contribution to relic dark matter density, it is necessary to know their annihilation cross-sections into ordinary- as well as the superpartners. Such calculations have been made and restrictions on the parameter space have been placed by requiring the contribution of such particles to dark matter energy density to be in the range allowed by cosmological observations. Search in collider experiments in LEP 200, LHC, and Tevatron is on, but it will be several years before we have results.

5 Concluding remarks

In this brief review, I have not touched upon a multitude of other ideas and problems, particularly problems associated with *dark energy*. The enormous progress in observational cosmology and the unprecedented accuracy of the cosmological parameters have posed profound problems for both particle physics and cosmology. It is clear that the Standard Model of elementary particles and their interactions fails to provide a complete catalog of the building blocks of our Universe. Physics beyond the Standard Model, Grand Unified Theories, and supersymmetry have hints that they may provide the necessary ingredients, but it is far from clear. There is also the over-riding problem of baryon asymmetry. The symmetry between particles and antiparticles is firmly established in collider physics, yet there is no sign of that symmetry in the observed Universe. The observed Universe is composed almost entirely of matter with little or no primordial antimatter. There are various proposals to explain this asymmetry invoking violation of lepton number (L) during electro/weak phase transition (leptogenesis) or the violation of baryon number (baryon number-lepton number) during the phase transition at the grand unification scale (baryogenesis) [4]. There is no dearth of new ideas (extra spatial dimensions (large and small)), our universe a "Brane" in a multidimensional space and time, and so on. The inflationary Standard Model of cosmology has many problems of its own when it comes to details. Big questions remain to be answered. Did inflation occur at all? What is the origin of the hypothetical "inflaton" field that drove inflation? How did the different forms of matter/energy of comparable abundance become compatible with the transition to accelerated universe in the present epoch? What is the nature of the dark energy responsible for this accelerated expansion? In any case, the strong symbiotic relation between particle physics and astrophysics and cosmology has produced many new challenges.

Acknowledgements

This work was supported in part by the US Department of Energy (DOE) under contract no. DE-FG02-85ER40237. I am greatly indebted to Mark Trodden for many helpful discussions.

References

[1]   Wendy L. Freedman and Michael S. Turner, *Measuring and Understanding the Universe*, 2003, arXiv:astro-ph/0308418

[2]   Michael Turner, *The New Cosmology*, 2002, arXiv:astro-ph/0202007; Michel Turner, *Dark Matter and Dark Energy; The Critical Questions*, 2002 arXiv:astro-ph/0207297

[3]   Jonathan Feng, *Supersymmetry and Cosmology*, Slac Summer Institute, July 28–August 8, 2003, Stanford, California

[4]   Antonio Riotto and Mark Trodden, *Recent Progress in Baryogenesis*, Annu. Rev. Nucl. Part. Sci., 1999, **49**:35–75

Discussion

Q: M. DISNEY:
Is there any observation(s) that falsifies inflation?

A: K. C. WALI: .
As far as I know, WMAP observations have confirmed predictions based on inflationary hypothesis. There are some discrepancies, but they are matters of details. Maybe that will bring about better understanding of inflation.

Q: J.-C. PECKER:
Some "matter-anti-matter asymmetry" models have been tried and failed to fit observations [Alfvén, Omnès & Montmerle, Souriau].

A: K. W.:

I don't know these papers, but it is true from particle physics, so far there is no completely satisfactory theory to explain this asymmetry. Reference 4 in my paper summarizes the recent progress and current situation.

Q: F. SANCHEZ:

The pioneer of quantum cosmology was Eddington. He has predicted the "tau" with the right order of mass. He introduced also chiral symmetry, "Majorana algebra" and the nine-dimension space Clifford algebra. Do you know if someone is reconsidering Eddington's Fundamental Theory?

A: K. W.:

No. I don't know anyone who is working on Eddington's Fundamental Theory. I am surprised you say Eddington had predicted tau lepton.

Q: M. CASSÉ:

What is in your opinion the best way to discover the neutralino: (a) production in collider experiments, (b) passive detection in underground detectors, (c) indirect detection through gamma rays from their annihilation?

A: K. W.:

I am afraid I don't know the answer. Confirming evidence for the neutralino can only come from discovering some other particles predicted by supersymmetry. That suggests collider experiments.

Q: F. BERNARDEAU:

What are our best chances of detecting dark matter (colliders, direct detection, indirect detection...) depending on the flavor compositions?

A: K. W.:

I am afraid I don't know the answer.

Part IV

Large-scale structure

9

Observations of large-scale structure

Valérie de Lapparent

Institut d'Astrophysique de Paris, CNRS, Univ. Pierre et Marie Curie
98 bis Boulevard Arago, 75014 Paris, France

1 Introduction

Over the past decade, the observation of the galaxy distribution at large scale has made significant advances thanks to (i) the building of fiber spectrographs with a large field of view and a high multiplex gain (Lewis *et al.* 2002; Burles *et al.* 1999, Watson *et al.* 1998), and (ii) the dedication of large numbers of observing nights or the use of dedicated telescopes for such projects. These observations have led to extensive maps of the distribution of matter traced by the galaxies. The three major projects aimed at mapping the "local Universe" over large solid angles are:

• the 2dF Galaxy Redshift Survey, an Anglo-Australian collaboration;
• the Sloan Digital Sky Survey, a US–Japanese–German collaboration;
• the 6dF Galaxy Survey, another Anglo-Australian collaboration.

In the following, I review these surveys and the remarkable results that they have provided on the large-scale structure of the Universe. I also review the recent or undergoing surveys to higher depth.

2 The large solid angle surveys

2.1 The 2dF Galaxy Redshift Survey

The 2dF Galaxy Redshift Survey (2dFGRS) is now complete and covers $\sim$1 500 square degrees of the southern sky, distributed in two strip-like regions of 70 to 80° long in right ascension and $\sim$10° and $\sim$14° wide resp. in declination, plus $\sim$80 single fields dispersed over the Southern Galactic Cap. The photometric catalog is based on the APM catalog (for "Automatic Plate Measuring machine" used to scan the UK Schmidt photographic plates; Maddox *et al.* 1990), which has been re-calibrated using CCD images. The limiting magnitude of the 2dFGRS is $b_J = 19.45$. The spectroscopic observations are performed with the 2dF spectrograph (Lewis *et al.*

2002), installed on the Anglo-Australian 4 m telescope. The spectrograph has a 2° field (hence its name) and 400 fibers. For each spectroscopic field, an aperture plate is drilled using the accurate positions of the galaxies to be observed, and the 400 fibers are plugged into the holes by a robot, which allows a reliable identification of the spectra with the catalog objects. The complete 2dFGRS catalog provides reliable redshifts for 221 414 galaxies to $z \leq 0.25$ (Colless *et al.* 2003).

The redshift maps for the two strip-like regions in each galactic cap show that the alternation of walls and voids of galaxies, which was detected in the CfA redshift survey (de Lapparent *et al.* 1986) and the Las Campanas Redshift Survey (Shectman *et al.* 1996), extends out to $z \sim 0.1$. Many walls are tenuous, but some are dense and contain numerous fingers-of-God corresponding to groups and clusters of galaxies.

The most striking result of the 2dFGRS is the first-time detection of the coherent in-fall of galaxies onto the large-scale structures. By decomposition of the two-point correlation function along the line-of-sight and the transverse direction, one separates the different components contributing to the peculiar velocity field in the sample. Application to the 2dFGRS allows a good match by a two-component model describing (i) the random pairwise velocities (due to groups and clusters) and (ii) the coherent in-fall onto high-density regions. By fitting this model to the data, Hawkins *et al.* (2003) derive a constraint on the matter density: $\Omega_m^{0.6}/b \sim 0.47 \pm 0.08$, where Ω_m is the matter density parameter of the Universe, and b is the linear bias parameter for the galaxies (Blanton *et al.* 2000). Assuming a bias value $b \sim 1.0$ as also measured from the 2dFGRS (Verde *et al.* 2002; Lahav *et al.* 2002), this yields $\Omega_m \sim 0.27 \pm 0.06$, in good agreement with an independent estimate based on the *direct* measurement of peculiar velocities of galaxy pairs (Feldman *et al.* 2003).

Application of the group finding algorithm of Eke *et al.* (2004) shows the full hierarchy of structures in the 2dFGRS, which is expected in the gravitational insta-bility scenario. The maps of the groups detected by Eke *et al.* (2004) show that they densely populate the walls of galaxies, whereas the richest groups or clusters are rarer and tend to be clustered. The hierarchical structure can be quantitatively measured using the n-point correlation functions. If one excludes the dense super-clusters present in the 2dFGRS (because they introduce a bias in the mean density), the average n-point correlation functions scale as the $(n-1)$th power of the two-point correlation function, up to $n = 6$ (Baugh *et al.* 2004; Croton *et al.* 2004); such a behavior is expected if the large-scale structure forms by gravitational instability.

By application of an objective algorithm for detection of the voids in the 2dFGRS, Hoyle & Vogeley (2004) identify 289 voids with an average effective radius of $\sim$15 h^{-1} Mpc (with a Hubble constant $H_0 = 100\,h$ km s^{-1} Mpc^{-1}); these voids are nearly empty, with an average density contrast $\delta\rho/\rho = -0.94 \pm 0.02$, and their total volume corresponds to 40% of the volume of the Universe sampled by the survey. These results are in good agreement with the semi-analytical cold dark matter models of Baugh *et al.* (2004), which include feedback from supernovae.

2.2 The Sloan Digital Sky Survey

The Sloan Digital Sky Survey (SDSS) is an on-going project performed with a dedicated 2.5 m telescope built at the Apache Observatory (New Mexico, USA). The survey aims at obtaining both CDD spectroscopy and redshifts for $\sim$900 000 galaxies (with $z \leq 0.25$) and $\sim$100 000 QSOs (with $z \leq 3$) over $\sim$7 000 square degrees of the sky (i.e., 1/6 of the celestial sphere). In contrast to the 2dFGRS, the input catalog based on digital detectors guarantees a high accuracy and homogeneity of the database, a fainter limiting magnitude $r \leq 22.5$, and provides a larger number of filters (*ugriz*, Fukugita *et al.* 1996). Both the imaging camera and the spectrograph cover a three degree field of view, and the spectra are obtained with a dual spectrograph having a total of 640 fibers (Burles *et al.* 1999). So far, 60% of the galaxy redshifts have been obtained. Plugging the fibers into the aperture plates is performed every day by two people who are fully occupied by this task; this manual procedure has been preferred over a robot, which would imply a higher financial cost.

The data sub-samples that are already available correspond essentially to declination strips located in the Southern and Northern Galactic Caps. The northern strips are also common with the 2dFGRS, and show identical large-scale structure. The general distribution has the same features as in the 2dFGRS and previous surveys, with an alternation of sharp walls and nearly empty voids with diameters of 10 to 50 h^{-1} Mpc.

The power spectrum of the SDSS (Tegmark *et al.* 2004)) is not well-fit by a single power law and shows curvature at large scale ($\sim$100 h^{-1} Mpc). One of the advantages of measuring the galaxy power spectrum is that it allows one to significantly decrease the likelihood interval for the cosmological matter density measured from the Wilkinson Microwave Anisotropy Probe (WMAP) Bennett *et al.* 2003). The joint SDSS and WMAP measurements imply $h\Omega_m = 0.213 \pm 0.023$ (Tegmark *et al.* 2004), which, combined with the result of the Hubble Space Telescope Key Project for measuring the Hubble constant ($h = 0.72 \pm 0.08$), and the best fit WMAP baryon fraction $\Omega_b / \Omega_m = 0.17$, yield $\Omega_m = 0.30 \pm 0.03$.

The power spectrum for the 2dFGRS (Percival *et al.* 2001; Tegmark *et al.* 2002) shows similar behavior to that for the SDSS (Tegmark *et al.* 2004), with some dispersion comparable to that seen when comparing with the other existing galaxy surveys. This is symptomatic of the required assumptions about bias, redshift-space distortions, and non-linear evolution, which must be made when calculating a power spectrum (Peacock and Dodds 1994).

2.3 Topological analyses of the 2dFGRS and SDSS

The "sponge-like" nature of the galaxy distribution, in which galaxies lie in sharp walls and filaments alternating with voids, implies that the high-order moments of the distribution may play a discriminating role when comparing with model distributions. Calculating the n-point correlation functions, however, requires an accurate measurement of the mean density, which may be affected by the presence of rare superclusters in the considered sample, an effect that is frequently referred to as "departure from a fair sample of the Universe" (see Croton *et al.* 2004, for the effect of removing two superclusters from the 2dFGRS when calculating the n-point correlation functions).

Reliable constraints on the high-order moments of the galaxy distribution can be obtained indirectly using topological analyses. The common approach is to study the topological properties of the iso-density contours of the galaxy distribution considered as a point-process; the contours are obtained by smoothing the distribution with an appropriate window function and identifying the separating surface between the high and low density regions at some density threshold. The topological properties of the distribution can then be defined as the variations in the characteristic properties of the iso-density contours as a function of density threshold.

The integrated mean curvature, also called "genus," of the iso-density contours as a function of density threshold was calculated in 2-D (along the plane of the strip-like survey regions) for the 2dFGRS and SDSS by Hoyle *et al.* (2002a, b, resp.), and shows good agreement of both surveys with the Virgo Consortium Hubble volume ΛCDM simulations (Frenk *et al.* 2000). The SDSS 2-D genus analysis of Hoyle *et al.* (2002b) also shows agreement with the analytical predictions from a Gaussian random field. The 3-D genus analysis of Hikage *et al.* (2002) based on the SDSS Early Data Release (EDR) also suggests better agreement with a Λ-dominated spatially flat cold dark matter model than with a standard cold dark matter model (with no cosmological constant). These measurements are, however, subject to substantial noise and await confirmation from the full SDSS survey, once completed.

A complete description of the topology of the galaxy distribution is obtained by measuring the four Minkowski functionals: the volume fraction, the total surface area, the integral mean curvature (i.e., genus), and the integral Gaussian curvature (i.e. Euler characteristic), as a function of density threshold. Application to the existing SDSS data by Hikage *et al.* (2003) shows remarkable agreement with the results from mock simulations based on a Λ-dominated spatially flat cold dark matter model.

2.4 Anisotropies in the 2dFGRS and SDSS

With their large volumes, the 2dFGRS and SDSS maps carry the hope of reaching fair samples of the galaxy distribution, i.e., to be representative samples of the general galaxy distribution. However, both surveys exhibit a marked anisotropy between the Northern and Southern Galactic Caps (NGC and SGC resp.). By smoothing the SDSS galaxy maps with a smoothing length of 10 h^{-1} Mpc, Einasto *et al.* (2003) extracted objectively the superclusters and measured their integrated luminosity; the same technique, applied with a smaller smoothing length of 0.8 h^{-1} Mpc, allows a spatial detection of the groups and clusters contained in the maps. The authors measure a significant anisotropy characterized by the most luminous clusters and superclusters in the NGC being a factor 2 more luminous than the corresponding systems in the SGC.

Similar effects are detected in the 2dFGRS, in the form of a deficiency in the K-band infrared counts obtained by cross-identification of the 2dFGRS galaxies with the 2MASS infrared survey (Cole *et al.* 2001): The K-band counts show a 30% deficiency at $z \leq 0.1$ in the SGC compared with the NGC (Frith *et al.* 2003). This effect is confirmed by a re-analysis of the optical counts in the region of the 2dFGRS, based on new CCD photometry that allows one to check the 2dFGRS photometric scale: these observations confirm a 30% deficiency at $z \leq 0.1$ in the 2dFGRS SGC *number-counts* at magnitudes brighter than ~ 17 (Busswell *et al.* 2003), which the authors also detect in the corrected Durham-UK Schmidt Telescope redshift survey (Ratcliffe *et al.* 1998). This under-density in the SGC is also visible in the 2dFGRS *redshift distribution* at $z \leq 0.1$, whereas at $z \geq 0.1$, the redshift distribution is restored to a common mean density with the NGC (Colless *et al.* 2003). Busswell *et al.* (2003) also find that this "Local Hole" persists over the full area of the APM catalogue (Maddox *et al.* 1990) with a 25% deficiency at $B \leq 17$, which suggests that the under-density extends over $\sim 300\ h^{-1}$ Mpc $\times$ 300 h^{-1} Mpc on the sky as well as $\sim 300\ h^{-1}$ Mpc in the redshift direction. A corresponding under-density is detected by De Propris *et al.* (2002) in the catalogued clusters contained in the SGC of the 2dFGRS. In their void catalog, Hoyle and Vogeley (2004) also detected that the SGC voids are more underdense than in the NGC.

The deeper ESO-Sculptor redshift survey (de Lapparent *et al.* 2004) also reveals the presence along the line-of-sight of a 200 h^{-1} Mpc under-density at $z \sim 0.37$ followed by a 100 h^{-1} Mpc over-density at $z \sim 0.45$ (see Fig. 9.1). Very-large-scale structures on a scale of $\sim 100\ h^{-1}$ Mpc have also been detected in the distribution of

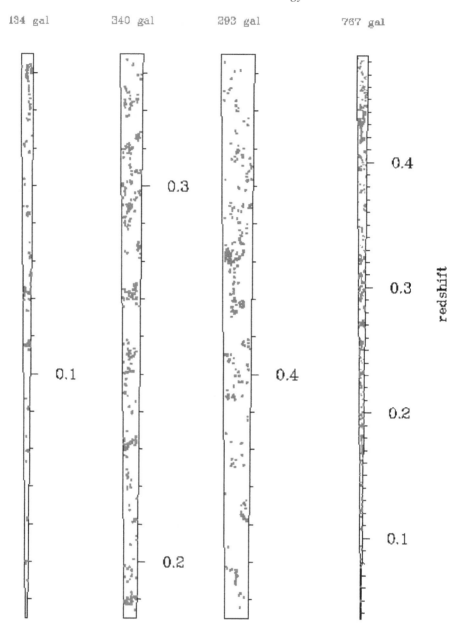

Figure 9.1 The galaxy distribution in the ESO-Sculptor redshift survey (de Lapparent *et al.* 2004) in the redshift range $0.035 \leq z \leq 0.485$ (right cone) and in a close-up view over three sub-intervals in redshift (the three left cones); the number of galaxies in each cone is indicated.

radio-galaxies at $z \sim 0.27$ (Brand *et al.* 2003) and in the distribution of quasars at $z \sim 1.2$ (Clowes and Campusano 1991). The latter structure is also associated with an excess of MgII absorbers (Williger *et al.* 2002) and most recently with an excess of passively evolving galaxies (Haines *et al.* 2004); a similar galaxy over-density was also detected at $z \sim 0.8$ by Haines *et al.* (2004).

Such anisotropies indicate that very large-scale structures exist in the galaxy distribution, which may reflect similar structure in the mass distribution, unless some systematic large-scale variations operate in the galaxy bias. The issue is how frequent these very large-scale structures are, and to which mass density contrast they correspond, as excess variance at large scale could call into question the current paradigm of Gaussian initial perturbations (see Miller *et al.* 2004).

2.5 The 6dF Galaxy Survey

Another redshift survey of the local universe (to $z \leq 0.1$) is the 6dF Galaxy Survey. It aims at measuring the redshifts of $\sim 150\,000$ galaxies over most of the southern sky, i.e., 17 046 square degrees. The input catalog is largely based on the 2MASS Extended Source Catalog (Skrutskie 2001, Jarrett *et al.* 2000), limited to all galaxies with infrared magnitude $K \leq 12.75$. The spectra are obtained using the 6 degree field (6dF) multi-fiber spectrograph (Watson *et al.* 1998, 2000) installed on the UK Schmidt Telescope, which can record 150 simultaneous spectra over the 5.7 degree field of the UK Schmidt. So far, one third of the spectra have been obtained (Jones *et al.* 2004). The specificity of this catalog is that it provides an unbiased sample of "normal" galaxies, as the infrared wavelengths at which the input catalog was obtained are mostly sensitive to the underlying old stellar component of galaxies, in contrast to the UV and optical wavelengths, which are more sensitive to present star formation and thus the gas content and the interactions between galaxies. The 6dFGS will thus provide a complementary view of the local universe to that provided by the SDSS and 2dFGRS.

3 Deep redshift surveys

3.1 Redshift maps at $z \sim 0.5$

In parallel with the large solid angle surveys (SDSS, 2dFGRS, and 6dFGS), redshift maps to $z \sim 0.5$ have also been obtained during the past few years (Small *et al.* 1997, Bellanger and de Lapparent 1995, Yee *et al.* 2000). These surveys show that the large-scale structure observed in the shallower surveys further extends out to $z \sim 0.5$, with apparently similar topological properties and characteristics scales (see Fig. 9.1). These deeper surveys are, however, limited in volume, with an angular extent of the order of 1 degree on the sky (corresponding to a transverse extent of $\sim 10\ h^{-1}$ Mpc), and their largest dimension lies along the line-of-sight. Therefore, these surveys cannot be used to derive statistical measures of the topological properties of the galaxy distribution at $z \sim 0.5$.

3.2 Up-coming redshift surveys at $z \sim 1$

The next step in increasing our knowledge of the galaxy distribution is to perform redshift surveys out to $z \sim 1$. Two such redshift surveys have recently been started. They benefit from the high multiplex gain of the new multi-slit spectrographs installed on 10-m class telescopes: VIMOS on one Very Large Telescope (VLT) unit of the European Southern Observatory (at Cerro Paranal in Chile), and DEIMOS on one Keck unit (at Mauna Kea in Hawaii).

3.2.1 The Deep Extragalactic Evolutionary Probe

The Deep Extragalactic Evolutionary Probe (DEEP2; Coil *et al.* 2004) aims at obtaining the redshifts of 65 000 galaxies with $z > 0.7$ with a 50% sampling rate to a limiting magnitude $R < 24.1$ (half of the galaxies to this limit will have a redshift measurement). The targets are four fields of $2° \times 0.5°$ on the sky, two of which overlap with the SDSS; each field will thus probe a region of $\sim 20 \times 80 \times 1000\ h^{-3}$ Mpc^3. Redshifts are obtained using the DEIMOS spectrograph (Faber *et al.* 2003) installed on one 10-m unit of the Keck facility, providing the simultaneous slit spectra of 200 objects. So far, half of the survey is completed and successfully detects galaxies in the redshift interval $0.7 < z < 1.4$ (Coil *et al.* 2004).

3.2.2 The VIMOS VLT Deep Survey

In parallel, the VIMOS VLT Deep Survey (VVDS; Le Fèvre *et al.* 2003b) aims at obtaining the redshifts of 100 000 galaxies with $I \leq 22.5$ over four fields of 4 square degrees each, thus reaching $z \leq 1.5$; each field will then probe a region of $\sim 80 \times 80 \times 2000\ h^{-3}$ Mpc^3. The redshifts are being obtained with the VIMOS spectrograph (Le Fèvre *et al.* 2003a) installed on one 8-m VLT unit, which allows one to obtain the simultaneous slit spectra of 400 objects. So far, 1/5 of the survey is completed.

4 Conclusions and prospects

The 2dFGRS and SDSS show the usefulness of mapping the galaxy distribution out to larger and larger distances over large areas of the sky. These two redshifts surveys have allowed important new advances in our understanding of the distribution of matter at large scale in the Universe: the detection and measurement of the coherent in-fall of galaxies onto the high-density regions; new measures of the power-spectrum of the galaxy distribution, which provide complementary constraints on the mass density parameter Ω_m to those provided by the recent cosmic microwave background measurements; better characterization of the high-order moments of the galaxy distribution and the related topological descriptions.

Nevertheless, despite their large size and their typical redshift depth of $z \sim 0.2$, the 2dFGRS and SDSS show a marked anisotropy between the Northern and Southern Galactic Caps, suggesting that these maps still do not represent fair samples of the Universe. Moreover, both surveys also yield controversial results on the nature of the galaxy bias (see, for example, Croton *et al.* 2004; Wild *et al.* 2004); the bias describes how the galaxies trace the underlying mass distribution, and is intimately related to the physical processes at play in the formation of large-scale structure. Deeper surveys to larger distances have been or are being performed to $z \sim 0.5$ and $z \sim 1$ resp., at the expense of angular coverage, thus providing only narrow pencil-beam probes of the galaxy distribution.

The present challenge in mapping the large-scale structure of the Universe is to obtain sufficiently densely-sampled and large-volume surveys out to $z \sim 1$. A significant gain in the statistical analyses of galaxy clustering could be obtained by such redshifts maps over $\sim$1000 square degrees of the sky, which would then probe a region of $\sim 500 \times 500 \times 1000\, h^{-3}$ Mpc^3; this would correspond to a surface area increased by factor of 100 over the current DEEP2 and VVDS surveys (see Section 3.2).

Such surveys would allow one to perform reliable statistical analyses of the topology of the distribution and to obtain better constraints on the size, density contrast and frequency of the very large-scale fluctuations. These various statistics would allow one to check the consistency with the fluctuation spectrum of the cosmic microwave background. Any disagreement between these two observational approaches would question the present concordance model based on the gravitational instability picture in a spatially flat Universe with a non-zero cosmological constant (Riess *et al.* 1998, Perlmutter *et al.* 1999, Phillips *et al.* 2001, Tonry *et al.* 2003).

Maps of the galaxy distribution at $z \sim 1$ would also allow one to put direct constraints on evolution in the galaxy clustering with redshift, by comparing the statistical properties measured at different redshifts. This would provide complementary observations to the local galaxy maps and the "distant" cosmic microwave background, to be also matched by the N-body models.

5 Acknowledgements

I am grateful to Jean-Claude Pecker for his kind invitation, which allowed me, while preparing this review, to revisit the wonders of the large-scale galaxy distribution. It was also a great pleasure to be able to give my paper inside the prestigious Collège de France.

References

Baugh, C. M., Croton, D. J., Gaztañaga, E., *et al.*, 2004, *MNRAS*, **351**, L44

Bellanger, C. & de Lapparent, V., 1995, *ApJ Lett.*, **455**, L103

Bennett, C. L., Hill, R. S., Hinshaw, G., *et al.*, 2003, *ApJS*, **148**, 97

Blanton, M., Cen, R., Ostriker, J. P., Strauss, M. A., & Tegmark, M., 2000, *ApJ*, **531**, 1

Brand, K., Rawlings, S., Hill, G. J., *et al.*, 2003, *MNRAS*, **344**, 283

Broadhurst, T. J., Ellis, R. S., Koo, D. C., & Szalay, A. S., 1990, *Nature*, **343**, 726

Burles, S., Pope, A., Uomoto, A., *et al.*, 1999, *Bull. Amer. Astron. Soc.*, **31**, 1501

Busswell, G. S., Shanks, T., Outram, P. J., *et al.*, 2004, *MNRAS*, **354**, 991

Clowes, R. G. & Campusano, L. E., 1991, *MNRAS*, **249**, 218

Coil, A. L., Davis, M., Madgwick, D. S., *et al.*, 2004, *ApJ*, **609**, 525

Cole, S., Norberg, P., Baugh, C. M., *et al.*, 2001, *MNRAS*, **326**, 255

Colless, M. M., Peterson, B. A., Jackson, C. A., *et al.*, 2003, astro-ph/0306581

Croton, D. J., Gaztañaga, E., Baugh, C. M., *et al.*, 2004, *MNRAS*, **352**, 1232

de Lapparent, V., Geller, M. J., & Huchra, J. P., 1986, *ApJ Lett.*, **302**, L1

de Lapparent, V., Arnouts, S., Galaz, G., & Bardelli, S., 2004, *A&A*, **422**, 841

De Propris, R., Couch, W. J., Colless, M., *et al.*, 2002, *MNRAS*, **329**, 87

Einasto, J., Hütsi, G., Einasto, M., *et al.*, 2003, *A&A*, **405**, 425

Eke, V. R., Baugh, C. M., Cole, S., *et al.*, 2004, *MNRAS*, **348**, 866

Faber, S. M., Phillips, A. C., Kibrick, R. I., *et al.*, 2003, in *Instrument Design and Performance for Optical/Infrared Ground-based Telescopes.* Edited by Iye, Masanori; Moorwood, Alan F. M. *Proceedings of the SPIE*, Volume 4841, 2003, 1657–1669

Feldman, H., Juszkiewicz, R., Ferreira, P., *et al.*, 2003, *ApJ Lett.*, **596**, L131

Frenk, C. S., Colberg, J. M., Couchman, H. M. P., *et al.*, 2000, astro-ph/0007362

Frith, W. J., Busswell, G. S., Fong, R., Metcalfe, N., & Shanks, T., 2003, *MNRAS*, **345**, 1049

Fukugita, M., Ichikawa, T., Gunn, J. E., *et al.*, 1996, *AJ*, **111**, 1748

Haines, C. P., Campusano, L. E., & Clowes, R. G., 2004, *A&A*, **421**, 157

Hawkins, E., Maddox, S., Cole, S., *et al.*, 2003, *MNRAS*, **346**, 78

Hikage, C., Schmalzing, J., Buchert, T., *et al.*, 2003, *PASJ*, **55**, 911

Hikage, C., Suto, Y., Kayo, I., *et al.*, 2002, *PASJ*, **54**, 707

Hoyle, F. & Vogeley, M. S., 2004, *ApJ*, **607**, 751

Hoyle, F., Vogeley, M. S., & Gott, J. R. I., 2002a, *ApJ*, **570**, 44

Hoyle, F., Vogeley, M. S., Gott, J. R. I., *et al.*, 2002b, *ApJ*, **580**, 663

Jarrett, T. H., Chester, T., Cutri, R., *et al.*, 2000, *AJ*, **119**, 2498

Jones, D. H., Saunders, W., Colless, M., *et al.*, 2004, astro-ph/0403501

Lahav, O., Bridle, S. L., Percival, W. J., *et al.*, 2002, *MNRAS*, **333**, 961

Le Fèvre, O., Saisse, M., Mancini, D., *et al.*, 2003a, in *Instrument Design and Performance for Optical/Infrared Ground-based Telescopes.* Edited by Iye, Masanori; Moorwood, Alan F. M. Proceedings of the SPIE, Volume 4841, pp. 1670–1681 (2003)., 1670–1681

Le Fèvre, O., Vettolani, G., Maccagni, D., *et al.*, 2003b, in *Discoveries and Research Prospects from 6- to 10-Meter-Class Telescopes II.* Edited by Guhathakurta, Puragra. *Proceedings of the SPIE*, Volume 4834, 2003, 173–182

Lewis, I. J., Cannon, R. D., Taylor, K., *et al.*, 2002, *MNRAS*, **333**, 279

Maddox, S. J., Efstathiou, G., Sutherland, W. J., & Loveday, J., 1990, *MNRAS*, **243**, 692

Miller, L., Croom, S. M., Boyle, B. J., *et al.*, 2004, *MNRAS*, **355**, 385

Peacock, J. A. & Dodds, S. J., 1994, *MNRAS*, **267**, 1020

Percival, W. J., Baugh, C. M., Bland-Hawthorn, J., *et al.*, 2001, *MNRAS*, **327**, 1297

Perlmutter, S., Aldering, G., Goldhaber, G., *et al.*, 1999, *ApJ*, **517**, 565

Phillips, J., Weinberg, D. H., Croft, R. A. C., *et al.*, 2001, *ApJ*, **560**, 15

Ratcliffe, A., Shanks, T., Parker, Q. A., *et al.*, 1998, *MNRAS*, **300**, 417

Riess, A. G., Filippenko, A. V., Challis, P., *et al.*, 1998, *AJ*, **116**, 1009

Shectman, S. A., Landy, S. D., Oemler, A., *et al.*, 1996, *ApJ*, **470**, 172

Skrutskie, M. F., 2001, *Bull. Amer. Astron. Soc.*, **33**, 827

Slezak, E. & de Lapparent, V., 2005, *A&A*, to be submitted

Small, T. A., Sargent, W. L. W., & Hamilton, D., 1997, *ApJS*, **111**, 1

Tegmark, M., Blanton, M. R., Strauss, M. A., *et al.*, 2004, *ApJ*, **606**, 702

Tegmark, M., Hamilton, A. J. S., & Xu, Y., 2002, *MNRAS*, **335**, 887

Tonry, J. L., Schmidt, B. P., Barris, B., *et al.*, 2003, *ApJ*, **594**, 1

Verde, L., Heavens, A. F., Percival, W. J., *et al.*, 2002, *MNRAS*, **335**, 432

Watson, F. G., Parker, Q. A., Bogatu, G., *et al.*, 2000, in *Proc. SPIE* **4008**, p. 123–128, *Optical and IR Telescope Instrumentation and Detectors*, Masanori Iye; Alan F. Moorwood; Eds., 123–128

Watson, F. G., Parker, Q. A., & Miziarski, S., 1998, in *Proc. SPIE* Vol. **3355**, p. 834–843, *Optical Astronomical Instrumentation*, Sandro D'Odorico; Ed., 834–843

Wild, V., Peacock, J. A., Lahav, O., *et al.*, 2005, *MNRAS*, **356**, 247

Williger, G. M., Campusano, L. E., Clowes, R. G., & Graham, M. J., 2002, *ApJ*, **578**, 708

Yee, H. K. C., Morris, S. L., Lin, H., *et al.*, 2000, *ApJS*, **129**, 475

Discussion

Q: J. NARLIKAR:

Is there any evidence in large-scale surveys for the kind of periodicity earlier found by Broadhurst *et al.* (1990) in pencil-beam surveys?

A: V. de L.:

No periodicity has been detected since then in other redshift surveys. However, the recent deep surveys (see Section 3.1) show a clear alternation of voids and walls on scales of 20 to 50 h^{-1} Mpc. In the ESO-Sculptor redshift survey, we measure an excess correlation in the spatial two-point correlation function at a scale of 25 h^{-1} Mpc (Slezak & de Lapparent 2005). This scale is significantly smaller than that quoted by Broadhurst *et al.* (1990), and there is no evidence for such a regularity as in a periodic signal.

Q: W. NAPIER:

It used to be claimed that there is a periodicity of structure on scales of $\sim$100 h^{-1} Mpc, but there was no sign of this in the power spectrum you showed. Have these claims gone away?

A: V. de L.:

If there was a periodicity in the galaxy distribution at a scale of $\sim$100 h^{-1} Mpc, it should indeed show up in the power spectrum. The absence of a feature at these scales agrees with the absence of periodicity in the deep redshift surveys performed recently or now under way (see Section 3).

Q: G. BURBIDGE:
Are there any areas of the sky that have been looked at by more than one group?

A: V. de L.:
The 2dFGRS and SDSS galaxy redshift survey have a region in common, which yields nearly identical large-scale structure. Note that multiple redshift measurements for a given galaxy obtained by separate groups using different telescopes and instruments systematically yield identical redshifts within the error bars.

Q: J. SURDEJ:
When going faint and to high redshifts, how much of the atmospheric absorption lines and sky background may affect and bias the detection and redshift measurement of the galaxies?

A: V. de L.:
At redshifts near 0.5 and beyond, the continuum of a galaxy spectrum represents at most a few percent of the sky background. The dominant atmospheric effects are the OH emission bands in the red part of the visible spectrum, which complicate the sky subtraction, and may act as emission lines in the galaxy spectrum if not cleanly removed. Because the intensity of these emission bands varies spatially and with time, their subtraction requires a delicate treatment. Spectroscopy at $0.5 \leq z \leq 1$ therefore requires *slit* spectrographs such as VIMOS (VLT) or DEIMOS (Keck Telescope) for sufficient sky sampling in the vicinity of the observed galaxy.

Q: J. SULENTIC:
You showed the 3-D distribution of galaxies revealed by the SDSS. Has anyone compared that distribution with the distribution of the quasars of similar and/or higher z?

A: V. de L.:
The quasars are much sparser tracers of the matter distribution than galaxies. There are therefore very few that lie inside a wide-angle galaxy redshift survey to redshift 0.2 such as the SDSS or 2dFGRS. Due to their narrow solid angle, the deeper surveys to $z \sim 0.5$ also contain no or few quasars. The large over-densities of quasar detected by Haines *et al.* (2004) lie at redshifts $0.2 \leq z \leq 2$, but are unfortunately *not* coincident with any of the existing or under-going deep redshift surveys. Note that the detected over-densities in the quasar distribution are so far compatible with Gaussian initial perturbations (Miller *et al.* 2004).

10

Reconstruction of large-scale peculiar velocity fields

Roya Mohayaee,[1] R. Brent Tully,[1,2] and Uriel Frisch[1]

[1] *Observatoire de la Côte d'Azur, B. P.4229, F-06304 Nice Cedex 4, France*
[2] *Institute for Astronomy, University of Hawaii, Honolulu, HI 96822, USA*

Abstract

A reconstruction method for recovering the initial conditions of the Universe starting from the present galaxy distribution is presented, which guarantees uniqueness of solutions. We show how our method can be used to obtain the peculiar velocities of a large number of galaxies, hence trace galaxies' orbits back in time and obtain the entire past dynamical history of the Universe above scales where multi-streaming has not occurred. When tested against a 128^3 ΛCDM simulation in a box of $200\,h^{-1}$ Mpc length, we obtain 60% exact reconstruction on scales above $6\,h^{-1}$Mpc. We apply our method to a real galaxy redshift catalog, the updated NBG (Nearby Galaxies), containing 1 483 groups, and clusters in a radius of 30 Mpc h^{-1}, and reconstruct the peculiar velocity fields in the local neighborhood. Our reconstructed distances are well matched to the observed values outside the collapsed regions if $\Omega_m(t) = 0.20 \exp(-0.26(t - 13))$, where t is the age of the Universe in Gyrs.

1 Introduction

Reconstruction of the initial condition of the Universe from the present distribution of the galaxies, brought to us by ever-more sophisticated redshift surveys, is an instance of the general class of *inverse problems* in physics. In cosmology this problem is frequently tackled in an empirical way by a *forward approach*. A *statistical* comparison between the outcome of an *N*-body simulation and the observational data is made, assuming that a suitable *bias* relation exists between the distribution of galaxies and that of dark matter. If the statistical test is satisfactory then the implication is that the initial condition assumed by the simulation is a viable one for our Universe, otherwise one changes the cosmological parameters until a statistical convergence between the observed and the simulated present Universe is achieved.

Since Newtonian gravity is time-reversible, one could integrate the equations of motion back in time and solve the reconstruction problem trivially if, in addition to their positions, the present velocities of the galaxies were also known. However, the peculiar velocities of only a few thousands of galaxies are known out of hundred of thousands whose redshifts have been measured. Thus, a second boundary condition, in addition to the present redshifts of the galaxies, has to be provided: As we go back in time the peculiar velocities of the galaxies vanish. Contrary to the forward approach where one solves an *initial-value problem*, in the reconstruction approach one is dealing with a *two-point boundary value problem* (in this case, only the functional dependence of one of the boundary conditions is given, namely: time $\rightarrow 0$ then peculiar velocities $\rightarrow 0$). In the former, one has a *unique* solution but in the latter this is not always the case.

The question remains whether unique reconstruction can be achieved. In this work, we report on a new method of reconstruction (Frisch *et al.* 2002, Mohayaee *et al.* 2004, Brenier *et al.* 1987) that guarantees uniqueness.

2 A brief review of previous approaches to reconstruction

The history of reconstruction goes back to the work of Peebles who traced the orbits of the members of the Local Group back in time (Peebles 1989). In his approach, reconstruction was solved as a variational problem. Instead of solving Newton's equations of motion, one searches for the stationary points of the corresponding Euler–Lagrange action. In his first work (Peebles 1989) only the minimum of the action was considered. Later on, it was found that when the trajectories corresponding to the saddle-point of the action were taken, a better agreement between predicted and observed velocities could be obtained for the galaxies in the Local Group (Peebles 1995). Thus, by adjusting the orbits until the predicted and observed velocities agreed, reasonable bounds on cosmological parameters were found (Peebles 1989) consistently favouring a low-density Universe ($\Omega_m = 0.1\text{--}0.2$; noteworthy at a time when there was a common preference for $\Omega_m = 1$).

Although rather successful (Shaya *et al.* 1995) when applied to catalogs such as NBG (Tully 1988) and also to mock catalogs (Branchini, Eldar, and Nusser 2002), reconstruction with such an aim, namely establishing bounds on cosmological parameters using measured peculiar velocities, cannot be applied to larger galaxy redshift surveys, which contain hundreds of thousands of galaxies for the majority of which the peculiar velocities are unknown. For large catalogs, the number of solutions becomes very large and not only is uniqueness completely lost but also one is never sure that the full solution space has been explored. In addition, numerical action-based codes are needed to solve the problem, which challenges current computer capacities.

A physical reason for multiple solutions is the collisionless nature of cold dark matter. Collisionless fluid elements can undergo multistreaming. Regions of multistream are bounded by *caustics* where the density is formally infinite and inside which the velocity field can have more than one value. This is a major obstacle to a unique reconstruction.

3 Monge–Ampère–Kantorovich (MAK) reconstruction

Reconstruction can be a well-posed problem for as long as we avoid multistream regions. The mathematical formulation of this problem is as follows (see Frisch *et al.* 2002, Mohayaee *et al.* 2004, and Brenier *et al.* 1987). Unlike most of the previous works on reconstruction where one studies the Euler–Lagrange action, we start from a constraint equation, namely the mass conservation,

$$\rho(\mathbf{x})\mathrm{d}\mathbf{x} = \rho_0(\mathbf{q})\mathrm{d}\mathbf{q} \tag{1}$$

where $\rho_0(\mathbf{q})$ is the density at the initial position, $\mathbf{q}$, and $\rho(\mathbf{x})$ is the density at the present position, $\mathbf{x}$, of the fluid element. The above mass conservation equation can be rearranged in the following form

$$\det\left[\frac{\partial q_i}{\partial x_j}\right] = \frac{\rho(\mathbf{x})}{\rho_0(\mathbf{q})}, \tag{2}$$

where det stands for determinant and $\rho_0(\mathbf{q})$ is constant. The right-hand side of the above expression is basically given by our boundary conditions: The final positions of the particles are known and the initial distribution is homogeneous, $\rho_0(\mathbf{q}) =$ const. To solve the equation, we make the following hypothesis: The Lagrangian map $(\mathbf{q} \to \mathbf{x})$, is the *gradient* of a *convex* potential Φ. That is

$$\mathbf{x}(\mathbf{q}, t) = \nabla_q \Phi(\mathbf{q}, t) \tag{3}$$

The convexity guarantees that a single Lagrangian position corresponds to a single Eulerian position, i.e., there has been no multistreaming.[1] These assumptions imply that the inverse map $\mathbf{x} \to \mathbf{q}$ also has a potential representation

$$\mathbf{q} = \nabla_{\mathbf{x}} \Theta(\mathbf{x}, t) \tag{4}$$

where the potential $\Theta(\mathbf{x})$ is also a convex function and is related to $\Phi(\mathbf{x})$ by the Legendre–Fenchel transform (e.g., Arnold 1978)

$$\Theta(\mathbf{x}) = \max_{\mathbf{q}}\left[\mathbf{q} \cdot \mathbf{x} - \Phi(\mathbf{q})\right] \quad ; \quad \Phi(\mathbf{q}) = \max_{\mathbf{x}}\left[\mathbf{x} \cdot \mathbf{q} - \Theta(\mathbf{x})\right] \tag{5}$$

---

[1] The gradient condition has been made in previous works (Bertschinger and Dekel 1989) on the reconstruction of the peculiar velocities of the galaxies using linear Lagrangian theory.

The inverse map is now substituted in Eq. (2) yielding

$$\det \left[\frac{\partial^2 \Theta(\mathbf{x}, t)}{\partial x_i \partial x_j} \right] = \frac{\rho(\mathbf{x})}{\rho_0(\mathbf{q})} \tag{6}$$

which is the well-known Monge–Ampère equation (Monge 1781, Ampère 1820). The solution to this 222-year-old problem has recently been discovered (Brenier 1987, Benamou and Brenier 2000) when it was realized that the map generated by the solution to the Monge–Ampère equation is the unique solution to an optimization problem. This is the Monge–Kantorovich mass transportation problem (Kantorovich 1942), in which one seeks the map $\mathbf{x} \to \mathbf{q}$, which minimizes the quadratic *cost* function

$$I = \int_{\mathbf{q}} \rho_0(\mathbf{q}) |\mathbf{x} - \mathbf{q}|^2 d^3 q = \int_{\mathbf{x}} \rho(\mathbf{x}) |\mathbf{x} - \mathbf{q}|^2 d^3 x \tag{7}$$

A sketch of the proof is as follows. A small variation in the cost function yields

$$\delta I = \int_{\mathbf{x}} [2\rho(\mathbf{x})(\mathbf{x} - \mathbf{q}) \cdot \delta \mathbf{x}] \, d^3 x \tag{8}$$

which must be supplemented by the condition

$$\nabla_{\mathbf{x}} \cdot (\rho(\mathbf{x}) \delta \mathbf{x}) = 0 \tag{9}$$

which expresses the constraint that the Eulerian density remains unchanged. The vanishing of δI should then hold for all $\mathbf{x} - \mathbf{q}$ that are orthogonal (in L^2) to functions of zero divergence. These are clearly gradients. Hence $\mathbf{x} - \mathbf{q}(\mathbf{x})$ and thus $\mathbf{q}(\mathbf{x})$ is a gradient of a function of $\mathbf{x}$.

Discretizing the cost (Equation (7)) into equal mass units yields

$$I = \min_{j(\cdot)} \left(\sum_{i=1}^{N} \left(\mathbf{q}_{j(i)} - \mathbf{x}_i \right)^2 \right) \tag{10}$$

The formulation presented in Equation (10) is known as the *assignment problem*: Given N initial and N final entries one has to find the permutation that minimizes the quadratic cost function. The cost function is indeed the minimum of an Euler–Lagrange action for inertial particles formulated in suitable space and time coordinates (Croft and Gaztañaga 1997). If one were to solve the assignment problem (10) for N particles directly, one would need to search among $N!$ possible permutations for the one that would have the minimum cost. However, advanced assignment algorithms exist that reduce the complexity of the problem from factorial to polynomial (e.g., see Hénon 1995 and Bertsekas 1998. Furthermore Hénon's adaptation of sparse and dense algorithm suitable for cosmological problems has a complexity of less than $N^{2.5}$ and has been used extensively in Mohayaee *et al.* 2004).

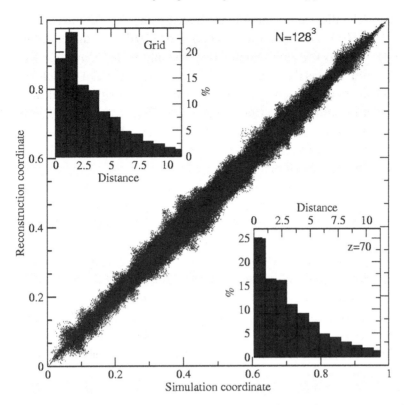

Figure 10.1 In the scatter plot, the dots near the diagonal are a scatter plot of reconstructed initial points versus simulation initial points for a grid of size 1.5 Mpc h^{-1} with more than 2 million points. The scatter diagram uses a *quasi-periodic projection* coordinate, $\tilde{\mathbf{q}} \equiv (q_x + \sqrt{2}q_y + \sqrt{3}q_z)/(1 + \sqrt{2} + \sqrt{3})$, which guarantees a one-to-one correspondence between $\tilde{q}$ values and points on the regular Lagrangian grid. The upper left inset is a histogram (by percentage) of distances in reconstruction mesh units between such points; the first bin corresponds to perfect reconstruction; the lower-inset is a similar histogram for reconstructed points at $z = 70$. The points at $z = 70$ are obtained by using Zel'dovich approximation to push particles back in time once their grid position has been reconstructed. Perfect reconstruction of about 18% is achieved in both histograms on scales of about 2 Mpc. On mesh sizes of about 6 Mpc h^{-1} this rate increases to about 60%.

4 Test against numerical simulation

We have tested our reconstruction against numerical N-body simulation. We ran a ΛCDM simulation of 128^3 dark matter particles, using the adaptive P^{3}M code HYDRA (Couchman *et al.* 1995). Our cosmological parameters are $\Omega_m = 0.3$, $\Omega_\Lambda = 0.7$, $h = 0.65$, $\sigma_8 = 0.9$, and a box size of $200\,\mathrm{Mpc\,h^{-1}}$. The simulations started at high redshift, in this case at $z = 70$. The results of our full box reconstruction are shown in Fig. 10.1. Once the assignment problem is solved the peculiar velocities can be simply evaluated using the Zel'dovich approximation $\dot{\mathbf{x}} = f(\Omega)H(t) \times (\mathbf{x} - \mathbf{q})$, where $f(\Omega) = \mathrm{d}\ln D / \mathrm{d}\ln a$ is dimensionless linear growth rate, $D(t)$ is the amplitude of the growing mode today, a is the cosmic scale factor, and $H(t)$ is the value of the Hubble parameter (Zel'dovich 1970). The peculiar velocities can then be used to reconstruct the positions $\mathbf{x}$ of the particles at any desired redshift back in time: $\mathbf{x}(z) = \mathbf{q} + (D(z)/D_0)(\mathbf{x}_0 - \mathbf{q})$, where $\mathbf{x}_0$ is their present positions, given by the simulation, and D_0 is the present value of D. The lower-inset of Fig. 10.1 shows the exact rate of reconstruction (when the separation between reconstructed and simulated positions of the particles is less than one mesh at $z = 70$) to be more or less the same as that of the top left inset. The reason is that particles move very little from the grid positions at high redshifts. However, a comparison between the two histograms demonstrates that yet another Zel'dovich approximation, which is involved in getting from grid positions to positions at $z = 70$, does not decrease the success of our reconstruction. (For detailed tests against simulations and reconstruction of statistics of the primordial density field, e.g., works on issues such as non-Gaussianity, see Mohayaee *et al.* 2005.) For scales below 2 Mpc corresponding to the smallest scale probed by reconstruction whose results are given in Fig. 10.1, the exact reconstruction rate is about 18% due to severe multistreaming at these scales. On larger scales of about 5 Mpc this rate increases to about 60%.

Outside collapsed regions the reconstructed peculiar velocities match well those simulated, as shown in Fig. 10.2. The primordial density field evaluated using these velocities also matches extremely well the simulated one as demonstrated in the lower panel of Fig. 10.2 (we thank S. Colombi for providing us with the lower panel of Fig. 10.2).

5 Application to real galaxy catalogs

We have applied our MAK method to the updated NBG catalog (Tully 1988), now including 3 300 galaxies within 3000 km s^{-1}. Other more extensive catalogs are available but this catalog provides good completion within the specified volume, which is sufficiently in depth for present purposes. The NBG has the important value-added feature of the detailed assignment of all objects to homogeneously identified groups and filamentary structures. The zone of Milky Way avoidance is shrinking as new surveys are integrated but before a dynamical model can be computed something must be done to account for galaxies lost due to obscuration. In this work, fake galaxies were created by reflection of objects at nearby higher latitudes in sufficient numbers to achieve the average density for the volume. Another correction to the catalog is one that accounts for incompleteness with distance. The correlation with mass is with the quantity of blue light. Light is lost from the catalog

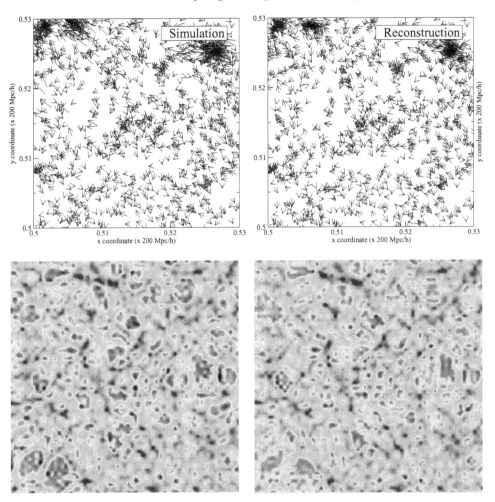

Figure 10.2 Top panel: Simulated (left panel) and the reconstructed (right panel) velocity field are shown for a thin 6 Mpc h^{-1} slice in x and y direction and full-box projection in the z direction of the simulation box. The reconstruction works extremely well outside dense/collapsed regions. Bottom panel: Simulated density field (left panel) and the reconstructed field (right panel) are shown for a thin slice cut of the simulation box. (Upper and lower panels do not correspond to the same slices of the simulation box.)

as galaxies become increasingly excluded with distance. Fortunately the problem is not extreme over the limited range of this study. Selection function corrections to luminosity range from unity at less than 10 Mpc (inside which there is completion because a low luminosity clip is imposed at $M_B = -16$) to only a factor 2.4 at 3000 km s^{-1}. The second observational component is a catalog of galaxy distances. In all, there are over 1 400 galaxies with distance measures within the

3000 km s^{-1} volume. In the present study, distances are averaged over groups because orbits cannot meaningfully be recovered on sub-group scales. The present NBG catalog is assembled into 1 234 groups (including groups of one) of which 633 have measured distances.

This catalog of galaxy positions, luminosities, and distances provides the basis for orbit reconstructions using MAK procedures. The distances, d, permit an extraction of peculiar velocities $V_{pec} = V_{gsr} - dH_0$, where V_{gsr} is the observed velocity of an object in the galactic standard of rest.

For MAK reconstruction the particles must all have the same mass since all the particles on the initial grid must be equal and each orbit reconstruction has equal weight. Consequently the endpoint elements must be broken up by different amounts depending on their supposed relative masses. In the first approximation of constant mass-to-light ratio M/L then the elements are simply broken into a number of particles that depends on L_i. The unit size is chosen to correspond to $10^9 L_\odot$, the faint end cutoff of the catalog. The elements are all located in redshift space (i.e., at their positions on the sky and at a distance inferred from their velocities). However the breakup into particles for the MAK reconstruction does not preserve the velocity distortion from real positions within elements; i.e., on sub-group scales. As we have demonstrated in the previous section, the MAK reconstructions of N-body simulations demonstrates good recovery of orbits on scales greater than 5 h^{-1} Mpc but clearly orbits cannot be recovered in shell-crossing regions.

The orbit of an element is defined by the center of mass of all the constituent particles as a function of time. The relationship between redshift and real space is estimated using the Zel'dovich approximation $\mathbf{v} = f(\Omega)(\mathbf{x} - \mathbf{q})$, where $\mathbf{v}$ is the peculiar velocity vector, $\mathbf{x}$ is the current Eulerian position, $\mathbf{q}$ is the initial Lagrangian position, and $f(\Omega) \sim 1 + b/\Omega_{m,0}^{4/7} + (1 + \Omega_{m,0}/2)\Omega_{\Lambda,0}/70$ and b is the bias factor, which we take equal to 1. In principle with our methodology, a variable bias can be obtained by varying M/L with location. In this discussion, the same M/L is assigned to all objects.

Once the particles are reconstituted into the catalog elements, a specific model defines positions that can be tested against observed positions. In Figs. 10.3, we show two MAK results. The left panel is the peculiar velocity field reconstructed by MAK of all the entries in the NBG catalog. There is a clear flow towards the great attractor as expected. The right panel shows a scatter plot of reconstructed versus observed distance moduli, $\mu_i = 5\log d_i + 25$. The scatter is mainly due to poor reconstructions near big clusters such as Virgo. In the infall region of Virgo, one is in the highly non-linear regime and moreover in a triple-valued region due to redshift space distortion. In this region, velocities deviate significantly from Hubble flow and MAK reconstruction does not necessarily find the right solution.

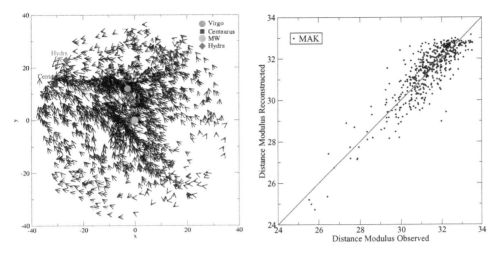

Figure 10.3 Left plot: Velocity field of objects in NBG catalog obtained by MAK reconstruction is shown. Large-scale flow towards the great attractor is visible, which overshadows infall into the Virgo cluster. The supergalactic coordinates x and y are used. Right panel: The scatter plot between MAK reconstructed distance modulus and that given by observations for the 663 objects with measured distances in this catalog.

(For reconstruction in the infall region, used for determination of mass of Virgo cluster, see Tully and Mohayaee 2004.)

The overall MAK reconstruction can be evaluated by a χ^2 estimator. We evaluate the median value for the χ_i^2; between measured and observed distance moduli

$$\chi_i^2 = (\mu_{observed} - \mu_{MAK})^2 / \epsilon_i \tag{11}$$

where ϵ_i is the error assigned to $\mu_{observed}$, which is the observed distance modulus of galaxy (or group or cluster) i in the catalog. Values of χ_i^2 can be determined for the 633 objects in the catalog with distance measures for a given choice of density parameter Ω_m and age t.

In this study we have only considered flat topologies. We assume that $\Omega_\Lambda = 1 - \Omega_m$, where Ω_Λ is a measure of the energy density of the Universe. With this constraint, there is a fixed relation between Ω_m, H_0, and the age of the Universe, t, such that if two of these parameters are specified then the third is defined: $h = (1/t)(2/3)(1/\sqrt{(1 - \Omega_m)})\log((1 + \sqrt{(1 - \Omega_m)})/\sqrt{(\Omega_m)})9.78$, where $h = H/100$.

Constraints on the parameter space (Ω_m, t) are summarized in Fig. 10.4. The two broad bands locate the 95% confidence limits provided by WMAP spatial fluctuation and SDSS power spectrum studies (Spergel *et al.* 2003; Tegmark *et al.* 2004). The heavy solid line indicates the locus of χ^2 minima as a function

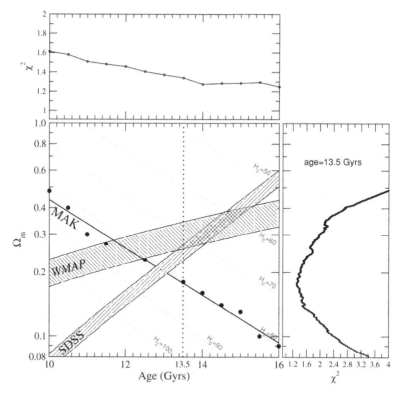

Figure 10.4 Constraints on the parameters Ω_m and age of the Universe given by MAK, WMAP, and SDSS. The 2σ constraints from WMAP and SDSS are given as shaded bands. The minimum of the χ^2 trough with the MAK reconstruction is given as the heavy solid line. Hubble constant contours are superposed as light lines. The two side panels illustrate aspects of χ^2 with the MAK reconstruction. In the top panel, the minimum value of χ^2 is shown at each age (i.e., at the location of the heavy solid line). In the right side panel, the values of χ^2 are shown for the range of Ω_m considered for the specific age $t = 13.5$ Gyr (i.e., the trace indicated by the vertical dotted line). It is seen that there is reasonable agreement between the three methodologies in the vicinity of $t = 12$–14 Gyrs, $h = 0.8$, and $\Omega_m = 0.2$–0.3.

of age from the MAK reconstructions. This line is described by the equation $\Omega_m(t) = 0.20 \exp(-0.26(t - 13))$ with age t in Gyr. The right panel illustrates the dependence of χ^2 values on Ω_m at the fixed age of $t = 13.5$ Gyr. The top panel shows the weak dependence of χ^2 on age at the χ^2-minimum trough defined by the heavy solid line. The overall minimum along this trough is reached at 17 Gyr. Overall with Fig. 10.4 two important points are to be noted. First, the uncertainties resulting from the MAK analysis are almost orthogonal to the WMAP and (especially) the SDSS constraints. Second, the three results intersect, resulting in concordance with the cosmological parameters $t = 13.2 \pm 0.8$ Gyr, $\Omega_m = 0.25 \pm 0.05$, and $H_0 = 77 \pm 5$.

In conclusion, we have demonstrated that our MAK reconstruction scheme guarantees uniqueness on large scales and can be applied to large data sets containing millions of objects. It is now being used with real data for the reconstruction of large-scale velocity fields. The method has been tested against numerical simulations and has been shown to recover the peculiar velocities of a large number of galaxies with a high success rate (taking the simulation dark matter particles to trace galaxies). We have also shown that MAK can be applied to real data and reconstructed peculiar velocity fields in the Local Supercluster. The best reconstruction fits obey the relationship $\Omega_m(t) = 0.20 \exp(-0.26(t - 13))$, where t is the age of the Universe in Gyrs. This fit intersects the WMAP and SDSS results within their 2σ uncertainties in the range t: 13–13.5 Gyrs, whence $\Omega_m = 0.2 - 0.3$.

Acknowledgements

Materials presented in Section 4 are parts of an ongoing collaboration with M. Hénon, S. Colombi, and H. Mathis. Materials presented in Section 5 are parts of collaborations with J. Peebles, S. Phelps, and E. Shaya. We also thank J. Colin, S. Matarrese, and A. Sobolevskii for discussions and comments. R. M. is supported by a European Marie Curie fellowship HPMF-CT2002-01532. B. T. is partially supported by the BQR program of the Observatoire de la Côte d'Azur.

References

Ampère A.-M., 1820, Mémoire concernant . . . l'intégration des équations aux différentielles partielles du premier et du second ordre, *Journal de L'École Royale Polytechnique*, **11**, 1

Arnold V. I., 1978, *Mathematical Methods of Classical Mechanics* (Springer, Berlin)

Benamou J.-D. & Brenier Y., 2000, The optimal time-continuous mass transport problem and its augmented Lagrangian numerical resolution, *Numer. Math.*, **84**, 375 (www.inria.fr/rrrt/rr-3356.html)

Bertschinger E. & Dekel A., 1989, Recovering the full velocity and density fields from large-scale redshift-distance samples, *Astrophys. J.*, **336**, L5

Bertsekas D. P., 1998, *Network Optimization: Continuous and Discrete Models* (Athena Scientific) {Auction algorithm also available at http://web.mit.edu/dimitrib/ www/auction.txt}

Branchini E., Eldar A., & Nusser A., 2002, Peculiar velocity reconstruction with fast action method: Tests on mock redshift surveys, *Mon. Not. R. Astron. Soc.*, **335**, 53

Brenier Y., 1987, Décomposition polaire et réarrangement monotone des champs de vecteurs, *C. R. Acad. Sci. Paris*, **305**, 805

Couchman H. M. P., Thomas P. A., & Pearce F. R., 1995, Hydra: An adaptive-mesh implementation of P^3M-SPH, *Astrophys. J.*, **452**, 797

Croft R. A. & Gaztañaga E., 1997, Reconstruction of cosmological density and velocity fields in the Lagrangian Zel'dovich approximation, *Mon. Not. R. Astron. Soc.*, **285**, 793

Frisch U., Matarrese S., Mohayaee R., & Sobolevskii A., 2002, A reconstruction of the initial conditions of the Universe by optimal mass transportation, *Nature*, **417**, 260

Mohayaee R., Frisch U., Matarrese S., & Sobolevskii A., 2003, Reconstruction of the primordial Universe by a Monge–Ampère–Kantorovich optimisation scheme, *Astron. & Astrophys.*, **406**, 393

Brenier Y., Frisch U., Hénon M., Loeper G., Matarrese S., Mohayaee R., & Sobolevskii A., 2003, Reconstruction of the early Universe as a convex optimization problem, *Mon. Not. R. Astron. Soc.*, **346**, 501

Hénon M., 1995, A mechanical model for the transportation problem, in *Compte Rendu de l'Academie des Sciences*, **321**, 741. A detailed version including an optimization algorithm is available at http://arXive.org/abs/math.OC/0209047

Kantorovich L., 1942, On the translocation of masses, *C. R. (Doklady) Acad. Sci. URSS (N. S.)*, **37**, 199

Mohayaee R. Mathis, H., Colombi, S., & Silk J., 2005, *astro-ph/0501217*, *Mon. Not. R. Astron. Soc.*, in press.

Monge G., 1781, Mémoire sur la théorie des déblais et remblais, *Hist. Acad. R. Sci. Paris*, **666**

Peebles P. J. E., 1989, Tracing galaxy orbits back in time, *Astrophys. J.*, **344**, L53

Peebles P. J. E., 1995, Mass of the Milky Way and redshifts of the nearby galaxies, *Astrophys. J.*, **449**, 52

Shaya E. J., Peebles P. J. E., & Tully R. B., 1995, Action principle solutions for galaxy motions within 3000 km/s, *Astrophys. J.*, **454**, 15

Peebles P. J. E., Phelps S. D., Shaya E. J., & R. B. Tully, 2001, Radial and transverse velocities of nearby galaxies, *Astrophys. J.*, **554**, 104

Spergel D. N. *et al.*, 2003, First-year Wilkinson Microwave Anisotropy Probe (WMAP) observations: Implications for inflation, *Astrophys. J. S.*, **148**, 175

Tegmark *et al.*, 2004, Cosmological parameters from SDSS and WMAP, *Astrophys. J.*, **607**, 655

Tully R. R. B., 1988, *Nearby Galaxies Catalog*, Cambridge University Press (Cambridge, UK)

Tully R. B. & Mohayaee, 2004, Action model of infall into the Virgo cluster, astro-ph/0404006, in Outskirts of Galaxy Clusters, the proceedings of IAU Colloquim No. 195, 2004

Zel'dovich Ya. B., 1970, Gravitational instability: An approximate theory for large density perturbations, *Astron. & Astrophys.*, **5**, 84

Discussion

Q: M. DISNEY:

To infer peculiar velocities you need to know very accurately the foreground absorption in our own Galaxy. Cannot this seriously disturb the dynamical inference?

A: R. M.:

Corrections for foreground absorption are required but this issue is not a major source of uncertainty. It is important to have an all-sky description of the galaxy distribution but it is not necessary to map the peculiar velocity field through heavily obscured regions.

Q: M. DISNEY:

You might be interested to incorporate the HIPASS/HIJASS catalog – a 21 cm survey of the whole sky?

A: R. M.:

Definitely. We already have acquired HIPASS, which covers the southern sky and information from it will be included in our next catalog.

Q: J.-C. PECKER:

You know only the radial velocities of the galaxies, not the proper motions. How do you take into account transverse velocities and collisions or induced ejections of matter?

A: R. M.:

Our reconstruction method gives us the three components of the peculiar velocities, we then evaluate the radial components and observe distances to the galaxies in our catalog (in this case updated NBG, *Catalog of Nearby Galaxies*, Tully 1988). The merger effects and/or loss of matter are not taken into account. This, however, we do not believe to be an issue on large scales (above 4 Mpc) where we consider our reconstruction to be applicable.

Q: J.-C. PECKER:

The computation assumes massive points to represent galaxies; as you do not assume collisions to play a role, the process is strictly reversible, therefore I am not surprised of the success of the computation!

A: R. M.:

True, we assume galaxies trace dark matter on large scales. Dark matter is assumed to be "cold." Neglecting collisional effects and dissipations, which seem to be the case as confirmed by simulations, observations and theory, enables us to trace galaxies' orbits (taken as mass tracers) back into the early Universe.

Q: J. SURDEJ:

How sensitive is your reconstruction method to uncertainties on your input data?

A: R. M.:

The reconstructed distances are sensitive to input total radial velocities, which are given by the catalog. There are certain errors in these measurements: (1) the radial component of the peculiar velocities leads to redshift space distortion; (2) there are triple-valued regions due entirely to projection effects in redshift space that do not exist in real space. Such additional complications do affect the success of reconstruction. However, we have an approximate formulation for redshift space reconstruction, using real catalog.

Part V

Alternative cosmologies

11

The quasi-steady-state cosmology

Jayant V. Narlikar

Chaire Internationale, College de France, Paris
And
Emeritus Professor, Inter-University Centre for
Astronomy and Astrophysics, Pune, India

Abstract

Reasons are given as to why the standard cosmology does not give an entirely satisfactory description of the Universe and why one needs to look for alternative cosmology. An alternative cosmology is presented in which matter creation takes place in mini-creation events at regular intervals and in response the Universe oscillates on a short-term period of $\sim$50 Gyr while it also has a steady (exponential) long-term expansion at a characteristic time scale of $\sim$1000 Gyr. The explanation of the major observed features of the Universe in terms of this cosmology is given and new observations distinguishing it from standard cosmology are proposed.

1 Introduction

Any proposal to describe the Universe in terms different from the so-called standard cosmology is met with the criticism that, "If the standard model is working so well and now it is possible to quantify that model with great precision, why look for an alternative?." Before describing the quasi-steady-state cosmology (QSSC in brief) I will therefore spend some time in pointing out the weaknesses of standard cosmology, weaknesses that rob it of many of its merits as a scientific theory. First let me talk of the three claimed successes of standard cosmology.

The big-bang cosmology began with the advantage that the models predicting expansion of the Universe by Friedmann (1922, 1924) and Lemaitre (1927) came before the discovery of the phenomenon of recession of galaxies and Hubble's law (1929). Thus one can say that as a scientific theory the big-bang cosmology made a prediction (namely, that the Universe is expanding) that was successfully verified.

The second success claimed by standard cosmology is, however, of a mixed character. The early expectation of George Gamow was to be able to demonstrate that the origin of chemical elements was nucleosynthesis in the early Universe. This idea worked so far as light elements are concerned. Beyond the atomic weight 4, one needs to look towards other astrophysical processes, namely inside stars, for explaining the origin and abundances of most other nuclei. As argued by Geoffrey Burbidge (2005) in this conference, even the making of light nuclei in the big-bang nucleosynthesis (BBN) demands a rather finely tuned relation of the kind $\rho = \eta T^3$, with the value of the parameter η put in by hand. Moreover, as Burbidge and Hoyle (1998) have argued, alternative astrophysical scenarios are now known that could account for even the light nuclei.

The third and most visible success attributed to standard cosmology was the prediction of the cosmic microwave background radiation (CMBR) by Alpher and Herman (1948) and its subsequent finding by Penzias and Wilson (1965). Again, Burbidge (2005) in this conference has given the historical perspective, which brings up the prior finding of this radiation by McKeller (1941), and the rather strange coincidence that if all the helium were made in stars, the starlight resulting from such a process would have a thermalization temperature very close to the actual temperature of the CMBR today. This coincidence remains unexplained in the standard model. For a discussion of this coincidence, see Hoyle, *et al.* (2000).

The standard model since 1965 has acquired an image of being the right theory of the Universe, despite the fact that it has had to be modified several times since then. I shall refer to these modifications as "epicycles" in the classical Greek tradition.

The first major epicycle was introduced in 1981 through the concept of inflation, i.e., rapid exponential expansion of the Universe for a very short time ($\sim 10^{-36}$ seconds) when its linear size grew by a factor in excess of 10^{50}. This idea was needed to get rid of the fundamental problems of an initial space-time singularity, very small particle horizons, very large curvature, and the entropy problem. Although it is still not clear *which* of the several inflation ideas is the accepted one in terms of a well-established fundamental particle theory, the general belief seems to be to accord an uncritical acceptance to the phenomenon of inflation.

Inflation prompted another epicycle when linked to dark matter. The total density of the Universe must be equal to the critical density, if the concept of inflation is correct. The visible density of the matter is hardly a few percent of this value. There are indications of dark matter in considerably larger amounts than the visible matter, if the Newton–Einstein gravity theory is correct and provided most clusters are dynamically relaxed. So it became necessary to postulate dark matter in quantities large enough to make up the closure density, even though there was no observational support for it.

The next epicycle came when this density was found to be incompatible with the requirements of the BBN. It reduced the expected abundance of deuterium to nearly zero. To sustain the BBN therefore it was necessary that the bulk of the matter was declared to be "non-baryonic." Although there are no observations to date either in the labs or in the cosmos to directly indicate the existence of non-baryonic dark matter (NBDM), it is now accepted uncritically.

Moreover, NBDM is needed to explain why the microwave background is homogeneous at least at the level of 10^{-5}, for temperature fluctuations, despite the inhomogeneities of matter in the form of galaxies. In standard cosmology, matter and radiation were fully interacting in the early stages and so any inhomogeneities of one would be shared by the other. Since NBDM does not react with radiation, this problem is solved. However, further epicycles are needed on the nature of NBDM, whether it is hot (HDM) or cold (CDM) or mixed (MDM) and how it is distributed in relation to visible matter, which is specified by a biasing parameter. Additionally, structure formation theories bring their own epicycles like the transfer function, the assumption of percolation, etc.

The most recent epicycle is paradoxically the very first one used in cosmology, namely that known as the *cosmological constant*. In 1917, Einstein introduced this constant λ into general relativity in order to get a static model of the Universe, since in those days the concept of an expanding Universe was not known. When Hubble's observations became established, Einstein was the first to abandon this constant and revert to the original general relativity. Other cosmologists had from time to time dabbled in the usage of this constant whenever they felt that the observations demanded it. However, as has happened frequently, observational errors often turned out to have been underestimated and the need to have the constant diminished after a time. As late as 1997, most cosmologists did not feel that the constant was needed.

This situation changed dramatically with Type Ia supernovae. These are regarded as standard candles in the determination of distances of far-away galaxies, with redshifts as high as ∼1 or more. The distant supernovae seemed fainter than expected if the standard models without the cosmological constant were used. So the constant (λ or Λ) was brought into the picture. However, it now appears that a fixed (i.e., constant) λ is not sufficient to understand the data. Today cosmologists talk of a *variable cosmological constant*, and a Universe that changed from deceleration to acceleration in its expansion, because the supernova data so demand.

The magnitude of the cosmological constant, *if it is a constant*, posed a problem first highlighted by Weinberg (1989). If it is assumed that it arose out of inflation, through the phase transition from "false" to "true" vacuum, then its magnitude is too high compared with what is required by the present-day observations. The factor by which it needs to be reduced is as low as 10^{-108}. Thus one needs fine tuning of unacceptably high order.

Going back to inflation, therefore, now it is proposed that there is *today* a redistribution of visible matter, NBDM, and Λ. The last arrival on the scene takes up nearly 72% of the closure energy, the NBDM is relegated to second place at ∼24% while the visible matter that the astronomer sees accounts for only ∼4%. In other words, cosmology as per the standard model does not put much stress on what you see but on *what you do not see*.

Perhaps I have been too harsh in passing judgement on a collective exercise that some of the greatest intellects in science are participating in today. But the exercise seems to me to be far more speculative than any scientific theory demands. Certainly I do not see any justification for the phrase "precision cosmology" prevalent today, suggesting that the cosmological model is more or less well determined. Another popular phrase (indicative of complacency) is "concordance cosmology," wherein it is argued that now we know, more or less, all details of the Universe and how well they fit the overall standard paradigm.

One example will suffice to indicate the unease I feel at the way the situation is developing. When astrophysicists discovered neutron stars, they realized that the theory required the central density of such a star to be as high as 10^{15} times the density of water. Considerable work was done by nuclear physicists and astrophysicists together to understand the nature of such matter and its equation of state. In cosmology, at the time of inflation the density of matter was in excess of 10^{55} times the density of water. Yet no one seems to be worried about the state of this matter. At the more fundamental level, one may also ask what is the operational definition of measurement of time at 10^{-36} seconds.

Some of us feel that these issues are worrisome and one needs to address them in standard cosmology if one believes that therein lies the correct solution. On the other hand, one may also take the view that given these internal weaknesses of standard cosmology, searches for alternatives are not out of place. In any case supporters of the standard model often react to such criticism by asking: "Given that the standard model is wrong, do you have any alternative to offer?" It is in response to this question that I will now present an alternative approach to cosmology that was proposed by the late Fred Hoyle, Geoffrey Burbidge, and myself (see Hoyle, Burbidge, and Narlikar 1993). We refer to this cosmology as the *quasi-steady-state cosmology* (QSSC).

2 The quasi-steady-state cosmology: theory

In this cosmology, one begins with the proposal made by Victor Ambartsumian in the 1960s that the Universe provides evidence for explosive phenomena on various scales. Today we see these in quasars, active galactic nuclei, gamma ray bursts, etc. on the galactic and extragalactic scale. Additionally Ambartsumian (1961) felt that even the clusters of galaxies seem to indicate lack of equilibrium of the kind that suggests that they may be expanding from an initial explosive origin, an origin where new matter was appearing in the Universe. In standard cosmology it is assumed that the clusters are in dynamical equilibrium and to sustain that assumption non-baryonic dark matter is postulated. Today, there is indication of relaxation or equilibrium in only a few clusters. Thus the issue of whether the clusters as a whole are relaxed is still an open one.

Following Ambartsumian's ideas, Fred Hoyle, Geoffrey Burbidge, and I felt that a quantitative expression can be given to them in terms of the gravitation theory developed by Hoyle and I in 1964–66, based on Mach's principle (see Hoyle and Narlikar 1964, 1966), suitably extended to describe the creation of matter. Such a theory ultimately leads to equations like those of general relativity, together with (1) a *negative* cosmological constant and (2) a scalar field of negative energy and stresses to describe the creation of matter. In the usual notation these are given below:

$$R_{ik} - 1/2g_{ik}R + \lambda g_{ik} = -8\pi G\{T_{ik} - f(C_i C_k - 1/4g_{ik}C^m C_m)\} \qquad (1)$$

Here C is a scalar field of negative energy and pressure that describes the creation of new matter. (In standard cosmology, the space-time singularity denotes the instant when the whole Universe was created: Since the event is singular, one is permitted (?) to ignore the violation of the law of conservation of energy and momentum.) The new matter in the QSSC appears at the expense of the C-field. Thus there is overall conservation of energy and momentum in the Universe.

The λ-term in this theory is related to the rest of the matter in the Universe and is in fact negative in sign. Its magnitude is of the order of 10^{-56} cm^{-2}, which is of the right magnitude when considered in the context of modern cosmological observations. (Compare and contrast with the cosmological constant problem of the standard cosmology!) For details of the derivation of the field equations, see Hoyle *et al.* (1995).

The simplest assumption one could make about the Universe is that it is homogeneous and isotropic and that matter is created in it *also homogeneously*. This model is none other than the old steady-state model. However, it fails to give expression to the explosive creation events of the kind mentioned above. To describe them one needs to look at the equations *in a region of strong gravitational field*.

The situation in the new cosmology is the following. In general the creation of matter is in the form of Planck particles, which are particles of Planck mass (corresponding approximately to energy of 10^{19} GeV), which are unstable and decay into smaller particles like baryons and leptons. The creation occurs, however, only if the overall energy momentum of the creation field equals the threshold of momentum of the Planck particle. This condition is not normally satisfied at a typical point in space. However, in regions of strong gravitational field the background level of the scalar field can be raised high enough for creation to occur.

Take for example a massive collapsed object of mass M, spherically symmetric. One can show that if the C-field background is not strong the original Schwarzschild solution will provide a reasonable approximation to the actual solution in the neighborhood of M. In this case, the energy density of the C-field behaves as

$$C^m C_m \sim (\text{constant})/[1 - 2GM/r] \tag{2}$$

For the creation of a Planck particle of mass m_P, one must have (2) equalling m_P^2, the speed of light being taken as unity. This will only happen close to the Schwarzschild radius.

This is why matter is created only in pockets of strong gravitational field. Once it is created, it is also accompanied by a compensatory creation field. The latter being negative has a repulsive gravitational effect and so the created matter is ejected with large energy. Thus explosions are generated, without requiring conditions of space-time singularity as in the big bang. We term these events as *minibangs* or *minicreation events* and the massive objects, *creation centers*.

There is a feedback mechanism between these local events and cosmology: For the locally produced explosions expand space. In the next simplest model after the steady-state theory referred to earlier, the Universe oscillates about the steady-state solution. In such a solution, the scale factor S of the Universe expands and contracts with a shorter oscillatory time scale Q compared with the longer scale of steady expansion P. In a simplified version of the solution:

$$S = \exp[t/P] \times \{1 + \eta \cos(2\pi\theta(t)/Q)\} \tag{3}$$

where the parameter η is less than unity in magnitude and $\theta(t)$ is a monotonic function of t, which behaves almost as t, except close to the turning points of S. The mathematical and physical properties of such solutions has been described in detail by Sachs *et al.* (1996).

The feedback mechanism works this way. Consider the minimum of scale factor during a local oscillation. Since the energy density of the C-field varies as S^{-4}, it is maximum at this stage. This enables most creation events to work fully, thus creating new matter and also a C-field, the latter causing expansion to go fast. However, as S increases, the C-field drops in strength and the creation centers begin to work less and less efficiently. This slows down the expansion of the Universe and eventually the negative cosmological constant takes over and it begins to contract the Universe. However, during the contraction stage, the C-field strength rises and more and more creation centers come on line with the result that the contraction slows down and is ultimately reversed. Thus we have a complete cycle of period Q.

In such an oscillatory Universe the period Q may be as long as 50 Gyr, while the exponential time scale P is even longer at around 1000 Gyr. It is easy to see that because the magnitude of η is less than 1, the scale factor never becomes zero. The largest redshift z_{max} one sees in the present cycle is of the order 6–10. Although the Universe has an exponential expansion, each of its oscillations is physically identical to the previous one. This is because creation of matter takes place (in minibangs all over the Universe) when the scale factor is at its minimum, and the intensity of the creation field is highest. This new matter compensates for the density reduction that would otherwise have taken place due to exponential expansion. Likewise, even though dissipatory processes would have increased the entropy density, the same is brought down by the low entropy new matter.

The numbers quoted above are partly put in by hand and are partly related to the fundamental constants appearing in the theory, namely G, λ, and f. In all one can say that there are four independent parameters in the theory, P, Q, z_{max}, and the epoch t_0 at which we are observing the Universe. We will now see how the observable features of the Universe can be explained.

3 The QSSC: observations

A. The CMBR: It is interesting to see how the microwave background arises in this cosmology. It is the thermalized relic radiation left behind by stars that were born and that shone during all the previous cycles. For, as mentioned earlier, each cycle is identical with others and in each new matter is born and gets made into stars and evolves through normal processes of stellar evolution. Although the exponential expansion of the Universe prevents any occurrence of the classical Olbers Paradox, the question remains as to what happened to all the relic radiation left behind by stars in the previous cycles. As optical radiation it will have energy density far in excess of that in the normal night sky background. The answer to the question is that with the passage of time and physical processes mentioned below, this radiation gets thermalized and is seen as CMBR. For details of this discussion see Hoyle *et al.* (1994, 2000).

From the present stellar activity one can estimate the expected energy density of such radiation and its temperature on thermalization. The answer comes very close to 2.7 K at the present epoch! In fact, this is the same old "coincidence" referred to in Section 1, resurfacing here. However, we now see its relevance in terms of the thermalization of starlight from previous cycles. Thus it is no longer a coincidence. However, we need to know how the thermalization is carried out.

The thermalization is shown to be carried out by metallic whiskers that are naturally created and ejected by supernovae. This activity is at its most effective at the oscillatory minima of the scale factor S. Chandra Wickramasinghe (2005) at this conference will discuss the details of this process and the evidence for such whiskers. Also see Narlikar *et al.* (1997). The dust density required to thermalize starlight is of the order of 10^{-34} g cm^{-3}, well within the limit of the cosmic metal abundances.

One can also estimate the power spectrum of inhomogeneities since these arise from the latest thermalization of starlight (for details, see Narlikar *et al*. 2003). As this has cluster-wise inhomogeneity, the largest signal will be given by the angular distribution of clusters at redshifts of ~ 6–10. This turns out to give the peak at $l \sim 200$ commonly ascribed to Doppler shift at the last scattering in the standard cosmology. This is an instance of how the same observation may have a different explanation depending on the paradigm used.

Since iron whiskers get aligned by intergalactic magnetic fields, it is natural to expect some signature, however weak, to be found in terms of polarization of the CMBR. This is being estimated at present.

B. Light nuclear abundances: The light nuclear abundances can be explained in two modes in the QSSC. The minicreation events in the sense of high energy events are similar to the classical big bang *sans singularity*, of course. Thus one can show (see Hoyle *et al*. 1993) that light nuclei like deuterium or helium can be made in the minibangs. The density–temperature relation seems to be different, however, from that of the big bang and it demonstrates the non-uniqueness of the BBN process. Alternatively, so far as helium is concerned, one can get most of the abundance from relic helium made by stars from previous cycles. Since, over long time scales available to the QSSC, most contribution will come from low mass stars, this process does not increase the abundance of heavy elements, since these stars do not proceed to the stages of making carbon or heavier nuclei. As Burbidge and Hoyle (1998) pointed out, stellar processes can make all of the nuclei found in the Universe. Even deuterium can be made in processes like solar flares.

C. Magnitude-redshift relation for Type Ia supernovae: How does the QSSC cope with the observations of *m*–*z* relations for extragalactic supernovae of Type Ia? It may appear that since standard cosmology demands a large positive cosmological constant, the QSSC with a negative λ is bound to fail. However, this is not so as was demonstrated by Narlikar *et al*. (2002). The solution lies in the intergalactic dust in the form of whiskers that provide mild but significant dimming of distant supernovae. One can obtain the best-fit density of such dust, keeping it as a free parameter. The result is a value that lies very well in the range required to thermalize the starlight to produce CMBR.

It is, however, necessary to recall the history of this test in the period 1960–80, for galaxies, which ultimately resulted in the realization that other uncertain factors intervene to make the conclusions uncertain (Burbidge 2005). In the present case are we sure of the range of variation of the so-called standard candle? Are gravitational lensing events not introducing bias? Is there no evolution over redshifts exceeding 1? In any case, whether the QSSC is right or not, the impact of cosmic dust on this test needs to be carefully estimated: More observations are needed to find the nature and extent of intergalactic dust.

D. Dark matter: There is no compelling need in the QSSC for non-baryonic dark matter. If such matter exists, it may form part of the total contents of the Universe. By and large in a minicreation event, one expects conditions requiring the application of very high energy physics, such as GUTs, SUSY, etc. The difference from the big-bang scenario is that here such events are repeatable and so one can study them under observations just like an astronomer studies stars under observation in different evolutionary states.

The stars that have burnt out in the previous cycles will provide dark matter in the present cycle, since they are no longer shining. It is also likely that there are very faint but very old white dwarfs also forming part of dark matter.

E. Large scale structure: Work by Nayeri *et al.* (1999) has shown with the help of a toy model that there is a new way of approaching the problem of large-scale structure. The toy model was in the form of a computer simulation with the following protocol.

First take a unit cube in which randomly distribute $N(\sim 10^6)$ points, representing galaxies. Of these choose at random αN points, where

$$\alpha = \exp\{3Q/P\} - 1 \tag{4}$$

Around each chosen point "create" a new neighbor randomly within a pre-assigned distance of $\beta N^{1/3}$. Then expand the whole system homologously in all directions by a linear scale of $\exp[P/Q]$. This procedure ensures that "after expansion" the cube has the same number density of points. From this expanded cube, extract the inner cube of unit dimension having the same center and the same principal directions. Thus we now have a new unit cube with the same number of "galaxies," including a few newly created new neighbors. Repeat the procedure a few times and you see the emergence of clusters and voids. If we compute the two-point correlation function for the set of points in the cube, we discover that the distribution quickly (in 6–7 iterations) settles down to the observed $r^{-1.8}$ dependence for galaxies and clusters.

This suggests that creation of matter and its ejection may play a vital role in structure formation. The scenario that emerges is one in which a collapsed massive object in the creation center acquires new matter and grows, until the growth of the C-field makes it unstable and its exterior breaks apart and is ejected. Clearly work needs to be done to quantify the details of such a model.

4 Differences from the standard model

Narlikar and Padmanabhan (2001) have discussed the standard and QSSC models critically. They have stressed the need to work out further details of the QSSC model to the same level of sophistication as that which the standard model is worked at. Nevertheless, here are some clear differences between the two models.

A. Blueshifts: The QSSC predicts the existence of a population of very faint ($>27^m$) galaxies with blueshift not exceeding 0.1. These galaxies belong to the previous cycle close to the last maximum value of the scale factor S.

B. Very old objects: The existence of very old stars belonging to the previous cycles, e.g., very faint white dwarfs and stars of half a solar mass or lower that may have become giants, will be clear indications that the Universe is much older than what the standard model claims.

C. Baryonic matter: The existence of baryonic matter exceeding the limit tolerated by the standard model would be another distinguishing test.

D. Gravitational radiation: The minibangs are also expected to yield detectable gravitational radiation. Although the peak emission of these waves will not be at the optimum value for the present generation gravitational wave detectors, one does expect some signal from them. Narlikar and DasGupta (1993) have made tentative estimates, which need to be further focused. They also pointed out that the spectrum of gravitational wave background generated by these minibangs will be different from that generated by inflation. Future technology may be able to express judgement on this issue.

5 Concluding remarks

In the last analysis, theories and speculations have to be decided by facts. So the predictions of this alternative cosmology also deserve to be critically examined. To make them more focussed, additional work needs to be done, which requires more humanpower. This is hard to come by in the present climate wherein most cosmologists are disinclined to look at alternatives.

As will be clear, the QSSC does not express an opinion on the so-called anomalous redshift phenomena discussed in this meeting. It is possible to adapt it, however, to try to find theoretical frameworks to understand these mysterious phenomena. But this also requires more workers in the field and is thus a challenge for the future.

References

Alpher, R. A. and Herman, R. C., 1948, *Nature*, **162**, 774
Ambartsumian, V. A., 1961, *A. J.*, **66**, 536
Burbidge, G., 2005, *Proceedings of the Paris Colloquium*
Burbidge, G. and Hoyle, F., 1998, *Ap.J.*, **509L**, 1
Friedmann, A., 1922, *Z. Phys.*, **10**, 377
Friedmann, A., 1924, *Z. Phys.*, **21**, 326
Hoyle, F. and Narlikar, J. V., 1964, *Proc. Roy. Soc. A*, **282**, 191

Hoyle, F. and Narlikar, J. V., 1966, *Proc. Roy. Soc. A*, **294**, 138
Hoyle, F., Burbidge, G., and Narlikar, J. V., 1993, *Ap.J.*, **410**, 437
Hoyle, F., Burbidge, G., and Narlikar, J. V., 1994, *M.N.R.A.S.*, **267**, 1007
Hoyle, F., Burbidge, G., and Narlikar, J. V., 1995, *Proc. Roy. Soc. A*, **448**, 191
Hoyle, F., Burbidge, G., and Narlikar, J. V., 2000, *A Different Approach to Cosmology*, Cambridge University Press
Hubble, E. P., 1929, *Proc. Nat. Acad. Sci.*, **15**, 168
Lemaitre, G., 1927, *Ann. Soc. Sci. Bruxelles*, **XLVII A**, 49
McKeller, A., 1941, *Pub. Dom. Astrophys. Obs.*, **7**, 251
Narlikar, J. V. and DasGupta, P., 1993, *M.N.R.A.S.*, **264**, 489
Narlikar, J. V. and Padmanabhan, T., 2001, *Ann. Rev. Astron. & Astrophys.*, **39**, 311
Narlikar, J. V., Wickramasinghe, N. C., Sachs, R., and Hoyle, F., 1997, *Int. J. Mod. Phys. D*, **6**, 125
Narlikar, J. V., Vishwakarma, R. G., and Burbidge, G., 2002, *P.A.S.P.*, **114**, 1092
Narlikar, J. V., Vishwakarma, R. G., Hajian, A., Souradeep, T., Burbidge, G., and Hoyle, F., 2003, *Ap.J.*, **585**, 1
Nayeri, A., Engineer, S., Narlikar, J. V., and Hoyle, F., 1999, *Ap.J.*, **525**, 10
Penzias, A. A. and Wilson, R. W., 1965, *Ap.J.*, **142**, 419
Sachs, R., Narlikar, J. V., and Hoyle, F., 1996, *A.& A.*, **313**, 703
Weinberg, S., 1989, *Rev. Mod. Phys.*, **61**, 1
Wickramasinghe, N. C., 2005, *Proceedings of the Paris Colloquium*

Discussion

Comment :

F. SANCHEZ :

In your model, the Hubble ratio is variable, as in the big bang one. So, to explain the double large number correlation, you have to choose between:

(1) a variation of the "physical constants" involved, G (as Dirac), or m_p, or m_e, or h an d

(2) admitting we live in a particular epoch, rejoining the cosmologic interpretation of Dicke or the "anthropic principle" of Carter.

Q : J.-C. PECKER :

During the very dense minima of the life of the oscillating Universe, galaxies perhaps are there; uniformity is not there. How can you therefore justify the use of the homogeneous-isotropic assumptions implied by the solution of the GR equations?

A : J. V. N. :

The minimum scale factor phase is only about 200 times denser than the present density. Thus galaxies of previous cycles are able to survive. Moreover, the $\exp(t/P)$ part of the expansion slowly but surely wipes away inhomogeneities. A few years ago, Banerjee and I had demonstrated the stability of the quasi-steady-state solutions against small perturbations like density inhomogeneities, variations in creation rate, etc.

Q : M. MOLES :

A definitive aspect of the standard model is cosmic evolution, i.e., objects are younger at higher z-values. Some evidences has been produced about evolution of SFR (stellar formation rate), or other aspects, with z. Could you, Jayant, comment on these aspects within your QSSC?

A : J. V. N. :

In the QSSC, at any epoch, one should find a spectrum of ages for galaxies. Thus we may see very old galaxies from previous cycles with ages $< Q$, and also young galaxies born in this cycle. I believe some old galaxies at high redshift have been found. These cannot be understood in the standard model, but are quite possible in the QSSC.

Q : K. WALI :

What is the connection of this theory with the fundamental theory of matter – GUTs, supersymmetry, etc.? This theory may be a good theory of the observed Universe. But what about earlier states? Was the Universe always like this?

A : J. V. N. :

The mini-creation events are high-energy events and therein one expects the very-high-energy physics to operate, including GUTs, SUSY, etc. The difference between them and big bang is that these events have no space-time singularity and also they are repeatable.

The oscillatory part of the solution describes an evolutionary Universe but the physical conditions as a whole do not vary much since the scale factor may not alter by more than 30–50. Each cycle is, however, physically similar to the previous one.

Q : F. SANCHEZ :

Is the Hubble radius temporally invariant in your model? If you make this variable you would have to vary proton or/and electron mass to maintain the double large number correlations:

$$\hbar c / G\, m_e m_p = R_H / 2\, \lambda_p = (M_U/m_e)^{1/2} \qquad (\Omega = 1)$$

A : J. V. N. :

The constants P and Q are two time scales of the QSSC. The de Sitter type horizon is therefore of constant radius Pc, where c is the speed of light. This replaces the constant c/H of the old steady-state cosmology.

Q : A. BLANCHARD

I have been suprised by your arguments, because after having said that the standard big bang is failing, you did present a model that includes a long list of strange ingredients to my taste, just to reproduce the basic facts supporting the standard "big bang." My question is: "Is there any type of observation that is reasonably likely to occur in the next ten years, and that, if confirmed, will dismiss your theory in your eyes?"

A : J. V. N. :

The QSSC has only one "strange" ingredient, the negative energy scalar field. All other items are from standard physics. Now, to your question:

If extensive searches are made and no very old objects (age > 20 Gyr) – like low mass stars, globular clusters, white dwarfs, galaxies as a whole – are found then the QSSC would lose credibility in my eyes. I should mention that, in making this reply, I am denying myself the facility to use epicycles like those used by standard cosmology.

Q : F. SANCHEZ :

In the QSSC cosmology, what is the status of the famous "double great number correlation – or coincidence?"

A : J. V. N. :

The value of the cosmological constant in the QSSC is determined from the number of particles in the observable Universe. This number is of the order of 10^{80} (baryons) and it gives the "correct" value of the cosmological constant.

12

Evidence for iron whiskers in the Universe

N. C. Wickramasinghe

Cardiff Centre for Astrobiology, Cardiff University, 2 North Road, Cardiff, CF10, 3DY, UK

Abstract

We review evidence for the widespread occurrence of iron grains in the form of long slender whiskers of radii ~ 0.01 μm and lengths in the range ~ 5 μm and 1 mm in the Galaxy and in extragalactic sources. Such particles are characterized by their property of being able to thermalize starlight to much longer wavelengths than is possible with standard interstellar grains. The cosmological role of iron whiskers is briefly discussed.

1 Introduction

The existence of iron particles as a component of interstellar grains was first proposed by Schalén (1939), an iron composition being argued at the time by analogy with the composition of iron meteorites. Many years later we proposed that an iron component of grains may arise from the mass flows from protoplanetary nebulae, cool stars, and from the outflowing material of supernova explosions (Hoyle and Wickramasinghe 1968, 1970). Such a component was also shown to be consistent with data on the extinction curve of starlight (Wickramasinghe and Nandy 1972).

In our early models, however, the iron particles were regarded as being spherical or nearly spherical in shape, with radii typically ~ 0.01 μm. Particles in the form of slender whiskers were considered only much later to account for the high grain emissivities required in certain galactic infrared sources, and also as a possible contributor to the cosmic microwave background (Wickramasinghe *et al.* 1975, Edmunds and Wickramasinghe 1975). The extinction properties in the visual and ultraviolet waveband for iron whiskers would be nearly identical to those of spherical particles of the same radius, provided the whiskers were in random orientation.

2 Thermalization of starlight

In the late 1960s, following the discovery of the cosmic microwave background (Penzias and Wilson 1965), Fred Hoyle and the present author began to explore a variety of dust models that might lead to the thermalization of starlight into far infrared and microwave wavelengths. The close coincidence of the energy density of the 2.7 K background ($\sim 10^{-12}$ erg cm^{-3}) with the energy density of starlight in the Galaxy as well as the density of energy released from H to He conversion over a cosmological scale (Hoyle and Tayler 1964) gave an impetus for such a search. Whilst sub-micron sized spherical grains are woefully inadequate in their long-wave opacity values, graphite whiskers offered a partial solution, permitting thermalization to at least ~ 100 μm (Edmunds and Wickramasinghe 1975). Calculations for mixtures of graphite whiskers with standard interstellar grains are shown in Fig. 12.1 (Wickramasinghe and Wallis 1996).

After several unsuccessful attempts to find a more efficient long-wave thermalizer, we finally considered an astrophysically realistic model involving iron particles, but now in the form of long slender whiskers. To obtain optimal microwave absorption efficiencies, it turns out that we require a material (metal) with a conductivity value $\sigma \sim 10^{18}$ s^{-1}, which could be attributed to impure iron at cryogenic temperatures. The optical constants n, k are thus readily computed from classical electromagnetic theory giving $n \approx k \approx (\sigma\lambda/c)^{1/2}$, thus making it possible to calculate efficiency factors for extinction using Mie type formulae for infinite cylinders (Wickramasinghe, 1972). However, for wavelengths

$$\lambda \geq \frac{c}{4\sigma} \frac{1}{\ln(l/a)} \left(\frac{l}{a}\right)^2 = \lambda_m$$

these formulae cease to be strictly valid. For the case when $\lambda \gg \lambda_m$, however, the much simpler Rayleigh–Gans formulae for optical cross-sections can be used (Wickramasinghe, 1972). Wickramasinghe *et al.* (1992) considered a smooth transition between these two cases, and devised an algorithm for calculating optical cross-sections of iron whiskers of arbitrary length.

Whilst a wavelength independent conductivity was used in our original work on iron whiskers, a more realistic model of the complex refractive index of iron at cryogenic temperatures was later introduced using the Drude–Lorentz theory of metals. In this case the optical cross-sections of individual iron whiskers and hence the mass extinction coefficient of randomly oriented whiskers were calculated (Wickramasinghe and Hoyle 1994).

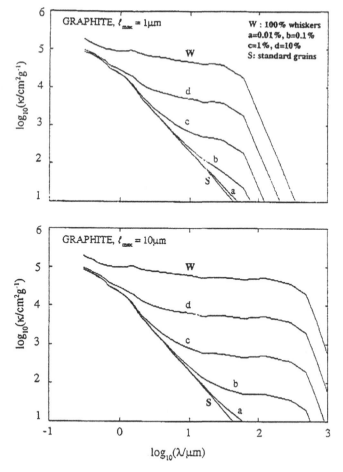

Figure 12.1 Mass absorption coefficient for mixtures of standard spherical sub-micron grains and iron whiskers. Curves marked W refer to graphite whiskers of radii $a = 0.01$ μm with lengths l distributed according to $n(l)dl \propto dl/dl$, $l > 0.02$ μm. Curves (a), (b), (c), (d) are for 0.01%, 0.1%, 1%, 10% mass contributions from graphite whiskers.

3 Iron whiskers from supernovae and elsewhere

The element iron is a major product of stellar nucleosynthesis and is expelled into interstellar space in the mass flows from supernovae. For an average Type II supernova a total mass of $\sim$0.1$M_\odot$ of iron is produced in an iron-rich shell. The shell, starting with a temperature of 10^{10} K and a density of 10^9 gm cm^{-3}, expands adiabatically to reach a temperature of 1000 K at a distance of 10^{16} cm. The ambient density of 10^9 cm^{-3} now permits the nucleation of iron particles to occur in a time scale of the order of months with growth of condensation nuclei proceeding to radii of $\sim$0.01 μm. Subsequent crystal growth is then presumed to continue preferentially

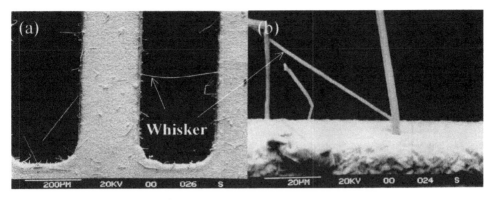

Figure 12.2 (a), (b) Tin whiskers growing in a computer circuit at the rate of ~1 mm per year.

in one direction only defined by the axis of a dislocation (Sears, 1957). Whisker lengths of ~1 mm would then be rapidly achieved owing to exponential growth, the whole process taking place under the conditions prevailing in a supernova shell within a time scale of ~10^7 s.

There are numerous laboratory studies to indicate that the growth of metal particles from a low density vapor leads to the formation of long slender whiskers (Sears, 1957; Nabarro and Jackson, 1958; Dittmar and Neumann 1958). It has been known to the chagrin of the computer industry that the growth of conducting metal whiskers often causes short circuits in certain types of tin-coated microcircuits. Fig. 12.2 shows an electron micrograph of whiskers bridging a gap between capacitor plates. The lengths of the whiskers are typically of the order of a millimeter and radii are on the general order of a few tenths of a micrometer. The precise mechanism for whisker formation on these surfaces still seems uncertain, but it has been suggested that growth occurs at nucleation sites on the surface from an overlying iron vapor. The growth of straight whiskers of diameters less than 1 μm and lengths of 1 mm is thus well documented. Furthermore it is known that growth to lengths of 1 mm (e.g., Fig. 12.2(b)) could take on the average 1 year. (See http://napp.nasa.gov/whisker/photos.)

4 Absorption by iron whiskers in the Galaxy

Figure 12.3 shows the mean mass absorption coefficients for a set of randomly oriented iron whiskers of diameter 0.02 μm with varying lengths. It is evident that the wavelength of the peak absorption increases from $\lambda \approx 100$ μm to $\lambda \approx 0.3$ mm with increasing particle length. Figure 12.4 shows the mean mass absorption coefficient for a mixture of two lengths, 100 μm and 1mm in the mass ratio 50:1. The top

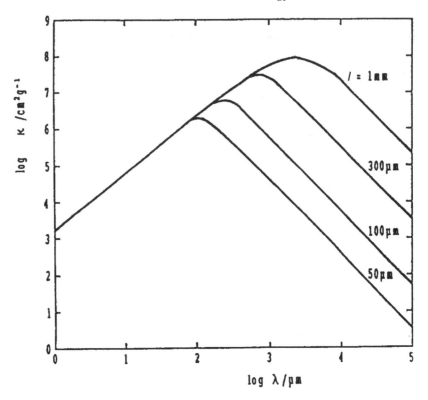

Figure 12.3 Mass absorption coefficient of randomly oriented iron whiskers of radii $a = 0.01$ μm and for various values of length, l.

scale on this plot refers to wavelengths of an emitting source redshifted to $z = 4.86$ corresponding to a thermalizing epoch placed at this value of z (Wickramasinghe and Hoyle 1994).

One of the earliest indications of iron whiskers in the galaxy came from observations of the emission spectrum of the Crab pulsar PSR0531 + 21 (Seward *et al.* 1985, Hoyle and Wickramasinghe 1988). Figure 12.5 shows this spectrum as a solid curve displaying a conspicuous dip over the frequency range 10^{11}–10^{12} Hz. This dramatic dip in flux can be elegantly explained on the basis of a shell of iron whiskers expanding outward at a speed of 3×10^7 cm s^{-1}. A total mass of iron whiskers of $\sim O_\Theta 1 M_\Theta$ distributed over a sphere of radius 10^{18}cm would then provide a column density of $\sim 10^{-5}$g cm^{-2}, which with $\kappa_{max} \sim 3 \times 10^6$ cm^2 g^{-1} yields an optical depth of order unity over the required waveband.

The presence of shorter iron whiskers has also been indicated by a deficit in the CII emission line intensity at 157.7 μm and in the NII line intensity at 205.3 μm towards the direction of the galactic centre (Nagakawa, *et al.* 1993, Bennet and

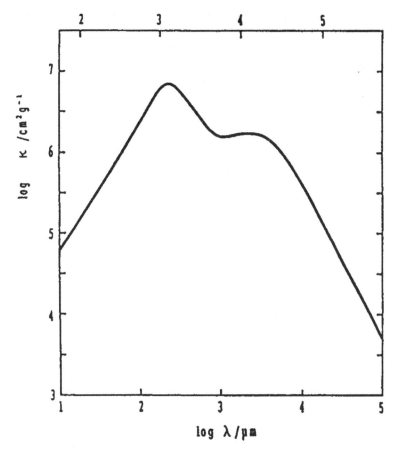

Figure 12.4 Absorption coefficient for a mixture of iron whiskers of lengths 100 μm and 1 mm in the mass ratio 50:1.

Hinshaw, 1993). Both these far-infrared emission lines have shown a significant decline of intensity in the galactic plane towards galactic longitude $l^{II} = 0$ compared with the fluxes at $l^{II} = -25°$. This is in sharp contrast with the behavior of the CO line at 2.6 mm, which does not diminish over this longitude range. The most reasonable inference is that there is a selective absorption of radiation in the 100–300 μm wavelength region. The possible role of grains with lengths in the range 50–100 μm is indicated as is evident in Fig. 12.6. It can be shown that $\sim 10^3 \, M_\Theta$ of iron whiskers arising from $\sim 10^3$ supernovae could exist in a volume of radius $\sim 300 \, pc$ around the galactic center (Wickramasinghe and Okuda 1993).

A more generally distributed population of iron whiskers was inferred by Dwek (2004b) from an interpretation of the infrared extinction towards the galactic center. Other localized sources of supernova-generated iron whiskers in the

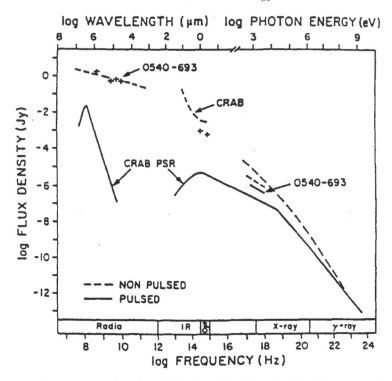

Figure 12.5 Spectrum of the Crab Nebula and Crab Pulsar showing a conspicuous dip in flux at $\sim 10^{11}$–10^{12} Hz (adapted from Seward *et al.*, 1985).

Galaxy are evident in the form of infrared excesses around supernova remnants, e.g., the Kepler SNR abd SNR CasA (Dwek 2004a, Morgan *et al.* 2003). Iron whiskers – on account of their relatively high values of mass absorption coefficient, $\kappa \approx 10^5$–$10^7 \, \text{cm}^2 \, \text{g}^{-1}$, are able to produce the observed infrared luminosities with modest masses of grains. Models involving classical grains, on the other hand, often violate abundance constraints with excessive dust masses being required.

5 Extragalactic evidence

The spectrum of the extragalactic supernova SN1987A at day 1300 displayed a behavior pattern strikingly similar to that of the Crab Nebula. The solid curve in Fig. 12.7 shows a calculation based on a model involving iron whisker emission in the vicinity of SN1987A and absorption by cold whiskers along the line of sight.

Narlikar *et al.* (1997) have pointed out that the emission spectra of radio quasars have a conspicuous dip in flux at a wavelength close to 1.3 mm. A sample of such spectra is reproduced in Fig. 12.8, from which the similarity to the behavior of the Crab pulsar (Fig. 12.5) is strikingly evident. The cores of quasars are presumably

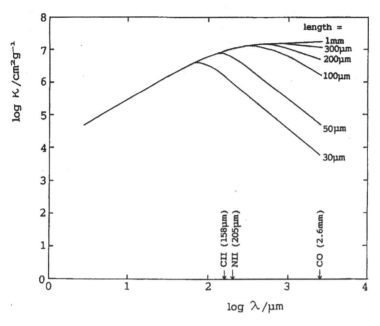

Figure 12.6 Mass absorption coefficient of randomly oriented iron whiskers of various lengths as functions of wavelength. Positions of CII, NII far IR lines, and the CO mm wave line are marked on the *x*-axis.

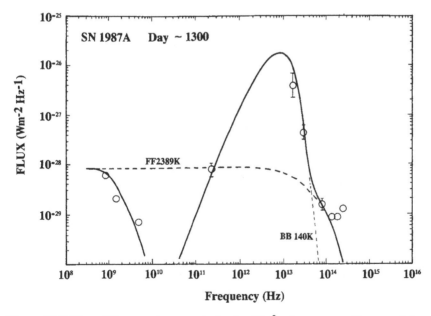

Figure 12.7 The solid curve shows emission by 10^{-5} solar masses of iron particles, plus en-route absorption by a large optical depth of 100 μm long iron whiskers. (For details see Wickramasinghe *et al.* 1992.)

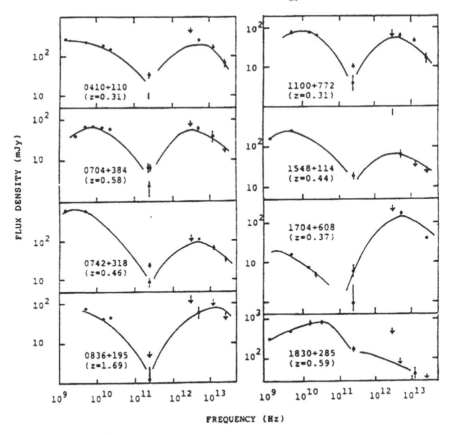

Figure 12.8 Flux curves for a sample of IRAS detected radio quasars showing a dip near λ ~ 1.3 mm. (Data from sources cited in Narlikar *et al.* 1997).

sites of violent star formation activity, so many supernova explosions might well contribute to an outflow of iron whiskers from these sources.

Observations of millimeter wave emission from several high redshift quasars could be modeled on the basis of localized clouds of iron whiskers. Whilst the longer millimeter length whiskers with larger values of κ are quickly expelled from the vicinity of sources owing to the effect of radiation pressure (terminal speeds $\propto \kappa^{1/2})$), shorter whiskers are slower moving and could provide longer-lived sources of circumstellar emission. The infrared emission arising from thermalization is then redshifted to the millimeter waveband due to Hubble expansion, an observed millimeter wave flux at a frequency v corresponding to emission at the wavelength $(c/v)/(1 + z)$, where z is the redshift. Spectral observations of two luminous, high redshift objects BR1202–0725 $(z = 4.69)$ and 4C41.17$(z = 3.8)$ were modeled on the assumption of iron whiskers with a mean length of 5 µm. In each case excellent agreement was obtained with $\sim 10^5 M_\Theta$ of such grains. Standard grains $\sim 10^8 M_\odot$ in the form of dust explain the same data.

6 Cosmological contribution

With ample evidence for iron whiskers, both near supernova sources and in surrounding interstellar regions, their wider role in contributing to the cosmic microwave background cannot be dismissed lightly. Details of any whisker-based explanation of the background, however, must depend on the cosmological model being considered. In the Quasi-Steady-State Cosmology proposed by Hoyle *et al.* (1993) the Universe undergoes cycles of expansion and contraction with a period typically of $Q \sim 40$ billion years, superposed on a more general cosmological expansion with time constant $P \sim 20Q$. The energy of the observed cosmic microwave background in this model is derived from starlight accumulated over many cycles. The thermalization is carried out in two stages. First, carbon whiskers thermalize the optical starlight into far infrared photons (see Fig. 12.1) at redshifts $z \sim 4.86$ when the Universe could have an optical depth of order unity from carbon whiskers, thus ensuring isotropy of the thermalized background. Next, iron whiskers (with an absorption curve given by Fig 12.5) take over, degrading the infrared radiation to yield the microwave background as observed.

The thermalized microwave background emanating from redshifts $z \sim 5$ would be initially unpolarized. A small degree of linear polarization, as has recently been observed, could, however, arise when this radiation passes through optically thin clouds of partially aligned iron whiskers, for example over the scale of clusters of galaxies. Whiskers with complete alignment in a particular direction would produce linear polarization close to 100%, but such an alignment is unrealistic. If the alignment arises owing to a process similar to that considered in theories of interstellar polarization, where grains spin rapidly in a magnetic field, the fractional alignment varies predominantly as B^2, where B is the magnetic field intensity. Since alignments and hence polarization of $\sim 1\%$ is achieved for interstellar grains in a mean magnetic field of 10^{-5} gauss, polarization to the extent of $10^{-4}\%$ will be achieved by iron whiskers in an intergalactic magnetic field of 10^{-7} gauss.

References

Bennett, C. L. and Hinshaw, G., 1993, in S. S. Holt and F. Verter (eds), *Back to the Galaxy*, New York, AIP

Dittmar, W. and Neumann, K., 1958, in R. H. Daramus, B. W. Roberts, and D. Turnbull (eds), *Growth and Perfection in Crystals*, New York, J. Wiley

Dwek, E., 2004, *Astrophys J.*, **607**, 848

Dwek, E., 2005, *Astrophys J.*, in press

Edmunds, M. G. and Wickramasinghe, N. C., 1975, *Nature*, **256**, 713

Hoyle, F. and Tayler, R. J., 1964, *Nature*, **203**, 1108

Hoyle, F. and Wickramasinghe, N. C., 1968, *Nature*, **217**, 415

Hoyle, F. and Wickramasinghe, N. C., 1970, *Nature*, **226**, 62

Hoyle, F. and Wickramasinghe, N. C., 1988, *ApSS*, **147**, 245

Hoyle, F., Burbidge, G., and Narlikar, J. V., 1993, *Astrophys J*, **410**, 437

Morgan, H. L. *et al*, 2003, *Astrophys. J.*, **597**, L33

Nabarro, F. R. N. and Jackson, P. J., 1958, in R. H. Daramus, B. W. Roberts, and
    D. Turnbull (eds), *Growth and Perfection in Crystals*, New York, J. Wiley

Nakagawa, T. *et al.*, 1993, in S. S. Holt and F. Verter (eds), *Back to the Galaxy*, New York,
    AIP

Narlikar, J. V., Wickramasinghe, N. C., Sachs, R., and Hoyle, F., 1997, *Int. J. Mod. Phys.
    D*, **6** (No.2), 125

Penzias, A. A. and Wilson, R. W., 1965, *Astrophys. J.*, **142**, 419

Schalen, C., 1939, *Uppsala Obs. Ann.*, **1**, No.2

Sears, G., 1957, *Ann. New York Acad. Sci.*, **65**, 388

Seward, F. D., Harnden, F. R., and Elsner, R. C., 1985, in M. C. Kafatos and R. B. Henry
    (eds), *The Crab Nebula and Related Supernova Remnants*, Cambridge, Cambridge
    University Press

Wickramasinghe, N. C., 1972, *Light Scattering Functions for Small Particles with
    Applications in Astronomy*, New York, Wiley

Wickramasinghe, N. C. and Hoyle, F., 1994, *ApSS*, **213**, 143

Wickramasinghe, N. C. and Nandy, K., 1972, *Rep. Prog. Phys.*, **35**, 157

Wickramasinghe, N. C. and Okuda, H., 1993, *ApSS*, **209**, 137

Wickramasinghe, N. C., Edmunds, M. G., Chitre, S. M., Narlikar, J. V., and Ramadurai,
    S., 1975, *ApSS*, **35**, L9

Wickramasinghe, N. C., Wickramasinghe, A. N., and Hoyle, F., 1992, *ApSS*, **193**,
    141

Wickramasinghe, N. C. and Wallis, D. H., 1996, *ApSS*, **240**, 157

Discussion

Q : B. LEMPEL :

1. Is there dust in the surroundings of SN (M1) (association with magnetism, MHD, jets, gravitation, and temperature)?

2. Does one observe dust at the end of the jets ejected from neutron stars or from the galactic nuclei?

A : C. W. :

1. Yes, dust is certainly observed in supernovae mainly through their infrared re-emission. This was certainly the case for SN1987A, and also in other instances.

2. Galactic nuclei also tend to be very dusty, and are strong emitters of infrared, again implying dust.

Q : M. DISNEY :

Surely the modern measurements of CBR, and in particular its isotropy and the exactness of fit to a BB spectrum, rule out its generation in any localized mechanism, such as iron whiskers.

A : C. W. :

This depends on the particular cosmological model one chooses. In the Quasi-Steady-State Cosmology of Hoyle, Burbidge, and Narlikar, the thermalization occurs at high redshifts, $z > 5$, and the starlight over several oscillatory cycles is supposed to contribute to the thermalized background.

Q : J.-C. PECKER :

I want to remind the audience of the excellent paper by Jean Lefèvre, published in *Annales d'Astrophysique* in the sixties. Lefèvre obtained from laboratory arcs in an inert atmosphere many iron whiskers, aligned by electric forces, collected, and observed in detail through an electronic micrcoscope.

A : C. W. :

Yes, that is certainly a very interesting paper you refer to. In fact whiskers, metal whiskers in particular, seem to be a fact of life in the Universe.

Q : J. SULENTIC :

I might suggest the quasars as a source of the iron whiskers. Circumstantial evidence includes:

1) About 60–70% of quasars show strong Fe emission lines (not to mention cold Fe via the Fe Kalpha line at 6.4 keV). Solar or supersolar abundances are typically found.
2) The same 60–70% show evidence for a strong wind or outflow (via blueshifted lines like CIV 1549).
3) The above are observed in quasars at all redshifts ($0 < z < 6$).

A : C. W. :

Yes, quasars are probably sites of active star formation, and supernova explosions. This would explain the high iron abundances in the cases you refer to. I think that whisker formation must surely accompany the explosions of supernovae in these objects, and the iron whiskers will be injected at high speed into the intergalactic medium along with the outflowing gas.

13

Alternatives to dark matter: MOND* + Mach

David Roscoe

Applied Maths, Sheffield University, Sheffield S3 7RH, UK

Abstract

Modified Newtonian dynamics (MOND) is an empirically motivated modification of Newtonian gravity (or, equivalently, of inertia) suggested by Milgrom as an alternative to cosmic dark matter. The basic idea is that at accelerations below $a_0 \approx 1.2 \times 10^{-10}$ m s^{-2} the effective gravitational attraction approaches $\sqrt{g_n a_0}$, where g_n is the usual Newtonian acceleration. This simple algorithm yields flat rotation curves for spiral galaxies and a mass-rotation velocity relation of the form $M \propto V^4$ that forms the basis for the observed luminosity-rotation velocity relation – the Tully–Fisher law.

The second approach, considered only very briefly here, is theoretically motivated and based on a hardline interpretation of Mach's principle.

1 Introduction

The appearance of discrepancies between the Newtonian dynamical mass and the directly observable mass in large astronomical systems has two possible explanations: Either these systems contain large quantities of unseen matter, or gravity on these scales is not described by Newtonian theory. Most attention has focused on the first of these explanations, the so-called CDM paradigm, which, whilst enjoying undeniable success cosmologically, also encounters severe observational difficulties within the context of the predicted distribution of dark matter in galactic systems (e.g., McGaugh and de Blok 1998a, Sellwood and Kosowsky 2001). There is no space here to discuss the possible problems with the CDM hypothesis and we merely comment that, as of 2004, candidate dark matter particles have not been detected by any means independent of their putative global

---

* This article is partly the article that Bob Sanders, who was unable to attend the Paris meeting, might have written and that part of it that deals explicitly with MOND is largely drawn from a substantial review article by Sanders and McGaugh (2002) and Figs. 13.1 and 13.3 are reproduced with their permission.

gravitational effect. It is this latter circumstance that justifies looking elsewhere for explanations of observed discrepancies in large gravitating systems.

To be credible, any alternative to dark matter should provide a more efficient description of the phenomenology and should make contact with familiar physical principles. To date, there are only two suggestions that begin to meet these requirements: The first (the one considered in most detail here) is Milgrom's empirically motivated MOdified Newtonian Dynamics, MOND (Milgrom 1983a,b,c,). The successes of the MOND hypothesis on scales ranging from dwarf spheroidal galaxies to super-clusters and its success (with the addition of certain reasonable assumptions) in predicting the acoustic peaks in WMAP data make it extremely difficult to ignore it as a viable alternative to the CDM paradigm.

The second (considered only very briefly here) is Roscoe's theoretical approach based upon a hardline interpretation of Mach's principle (Roscoe 2002,2004). Compared with MOND, this approach has had only limited development but, even so, it has enjoyed considerable success and must also be considered as currently viable.

2 Basis of MOND

2.1 An acceleration scale

The basis of MOND consists of two observational "facts" about spiral galaxies.

1. The rotation curves of spiral galaxies are asymptotically flat (Shostak 1973, Roberts and Whitehurst 1975, Bosma 1978, Rubin *et al.* 1980).
2. There is a well-defined relationship between the rotation velocity in spiral galaxies and the luminosity – the Tully–Fisher (TF) law (Tully and Fisher 1977, Aaronson *et al.* 1982), which implies a mass-velocity relationship of the form $M \propto V^\alpha$, where $\alpha \sim 4$.

As several authors over the years have noted, the first of these two facts can be accounted for in an ad hoc way by proposing that gravitational attraction becomes more like $1/r$ beyond some length scale that is comparable to the scale of galaxies. So the modified law of attraction about a point mass M would read

$$F = \frac{GM}{r^2} f(r/r_0)$$

where r_0 is a new constant of length on the order of a few kpc, and $f(x)$ is a function with the asymptotic behavior: $f(x) = 1$, where $x \ll 1$ and $f(x) = x$, where $x \gg 1$. However, it is easily seen that equating the centripetal to the gravitational acceleration for $r > r_0$ leads to the relation $v^2 = GM/r_0$, which, as Milgrom (1983a) realized, is incompatible with the observed TF law, $L \propto v^4$. Additionally, any modification attached to a length scale would imply that larger galaxies should exhibit a larger discrepancy (Sanders 1986) – but this is strongly contrary to the observations since low surface brightness spiral galaxies generally exhibit the largest discrepancies (McGaugh and de Blok 1998a).

Milgrom's brilliant insight was to recognize that the TF relation implied that any modifications of Newton's law should be tied to an *acceleration* scale rather than a *distance* scale. Thus, (skating over a few issues) Milgrom reasoned that in a *very weak* acceleration regime, the true gravitational acceleration $\mathbf{g}$ should be related to the Newtonian gravitational acceleration $\mathbf{g}_N$ as $\mathbf{g}^2 \sim \mathbf{g}_N$. Or, more formally, for the whole acceleration regime

$$\mathbf{g}\mu(|\mathbf{g}|/a_0) = \mathbf{g}_N$$

where a_0 is a new physical parameter with units of acceleration and $\mu(x)$ is a function that is unspecified but must have the asymptotic form $\mu(x) = x$ when $x \ll 1$ and $\mu(x) = 1$ when $x \gg 1$. Thus, in the weak-field regime, the magnitude of the effective gravitational force becomes: $g = \sqrt{g_n a_0}$. For a point mass M, setting g equal to the centripetal acceleration v^2/r gives

$$v^4 = GMa_0 \tag{1}$$

in the low acceleration regime. Thus, we obtain consistency with the original observational facts:

- All rotation curves of isolated masses are asymptotically flat.
- There is a mass-luminosity relation of the form $M \propto v^4$ consistent with the TF relation.

It is important to note that because MOND is based rigidly on the two observational "facts" alluded to at the beginning of the section then it is eminently falsifiable. For example, the unambiguous observation of a rotation curve of an isolated galaxy that declines in a Keplerian fashion at a large distance from the visible object would falsify MOND. Furthermore, it is important to note that MOND has only *one* free parameter – the mass-to-light ratio of the visible component of the object being modeled.

By contrast, it is very difficult to conceive of observations that would falsify the *multi-parameter* CDM paradigm and it is in this sense that one could argue that CDM is not a scientific hypothesis. See McGaugh and de Blok 1998b and McGaugh *et al.* 2000 for a discussion of various issues of this nature.

2.2 Subsequently confirmed general predictions of MOND

Apart from the startling success enjoyed by MOND in modeling the dynamics of LSBs in particular (see, for example, Fig. 13.1), there are several other direct observational consequences of the modified dynamics – all of which Milgrom

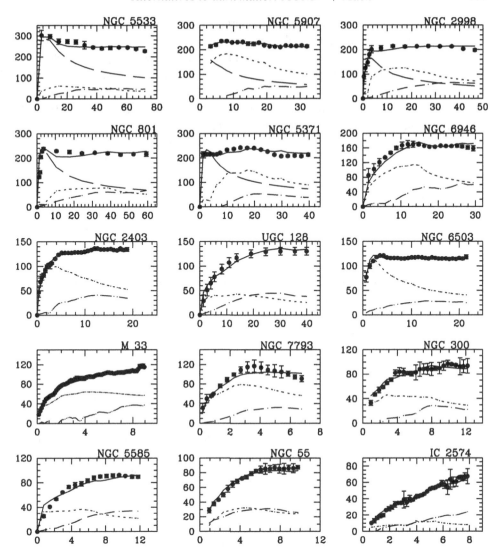

Figure 13.1 Rotation curves for a mix of LSB and HSB spirals. Filled circles are the observations, the solid lines are the MOND rotation curves whilst the dotted and dashed lines are the Newtonian rotation curves of the visible and gaseous components. Diagram by kind permission of Bob Sanders and Stacy McGaugh.

explored in his original papers – that are genuine predictions of MOND. They can be listed as follows.

1. There exists a critical value of the surface density $\Sigma_m \approx a_0/G$. If a system, such as a spiral galaxy, has a surface density of matter greater than Σ_m, then the internal accelerations are greater than a_0, so the system is always in the Newtonian regime. Thus, in HSB

galaxies ($\Sigma \geq \Sigma_m$) the visible mass should largely account for the observed dynamics according to Newton; that is, HSB galaxies should be well represented by "maximal disc" solutions (van Albada & Sancisi 1986). But in LSB galaxies ($\Sigma \ll \Sigma_m$, of which none had been observed when Milgrom made his original prediction) the visible mass would be incapable of accounting for the observed dynamics according to Newton. Milgrom's predictions have been amply verified – see, for example, McGaugh & de Blok (1998a,b).

2. MOND predicts the existence of a maximum surface density for spirals (see, for example, McGaugh *et al.* 1995b, McGaugh 1996) and this prediction is consistent with the observed upper limit on the mean surface brightness of spiral galaxies known as Freeman's law (Freeman 1970, Allen and Shu 1979). To conform with Freeman's law, such an upper limit must be put in by hand in dark matter theories (e.g., Dalcanton *et al.* 1997).

3. The optical disks of spiral galaxies with mean surface densities near this limit (HSBs) will be within an approximate Newtonian regime. Thus, according to MOND one would expect to see Keplerian fall-off to flatness in such rotation curves whilst, by contrast, it predicts that the rotation curves for LSBs ($\Sigma \ll \Sigma_m$) should rise continuously to flatness. This general qualitative difference between LSB and HSB rotation curves was first noted by Casertano and van Gorkom (1991). See Fig. 13.2 for typical examples.

4. According to Newtonian dynamics, near-isothermal pressure-supported systems have infinite mass. However, according to MOND such systems have finite mass with the density at large radii falling approximately as r^{-4} (Milgrom 1984). More particularly, according to MOND, a_0 appears as a characteristic acceleration in such systems and Σ_m appears as a characteristic upper-limit surface density (Milgrom 1984). Fish (1964) pointed out that elliptical galaxies exhibit a constant surface brightness within an effective radius. Subsequently, Corollo *et al.* (1997) showed that, within the general class of pressure-supported systems, there appears to be a characteristic surface brightness, which is on the order of that implied by Σ_m, i.e., the Fish law is recovered for the larger set of pressure-supported objects.

This list of genuine predictions that have been subsequently confirmed is to be compared with the total absence of corresponding predictions of the CDM model.

3 An alternative to MOND

It is well known that MOND is a purely phenomenologically derived algorithm for calculating the dynamics within astrophysical systems, and much effort has been put into searching for the "theory behind MOND" by Milgrom and collaborators – to date (2004) without success. An alternative approach, theory-driven and based entirely upon a particularly strong interpretation of Mach's principle, has been developed by this author (Roscoe 2002a, 2004). The successes of this approach, particularly in the modeling of LSB dynamics, raise the possibility that it might be the elusive "theory behind MOND."

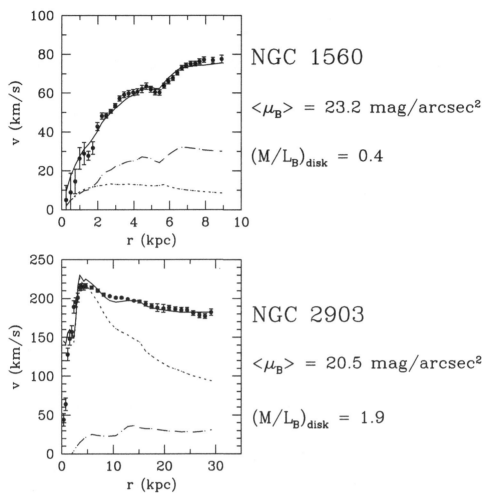

Figure 13.2 The filled circles show the observed 21-cm line rotation curves of a low surface brightness spiral, NGC1560, and a high surface brightness galaxy, NGC2903. The solid lines are the MOND rotation curves whilst the dotted and dashed lines are the Newtonian rotation curves of the visible and gaseous components. Diagram by kind permission of Bob Sanders and Stacy McGaugh.

3.1 The simple cosmology

The first paper (Roscoe 2002a) took the point of view that, since all our concepts of measurable "space and time" are irreducibly connected to the existence of material systems and to processes within such systems, then these concepts are, in essence, *metaphors* for the relationships that exist between the individual particles (whatever these might be) within these material systems. Since the most simple conception of physical space and time is that provided by inertial space and time, we were led to two simple questions:

Is it possible to associate a globally inertial space and time with a non-trivial global matter distribution and, if it is, what are the fundamental properties of this distribution?

In the context of the simple model analyzed, we were led to definitive answers to these questions so that:

- *A globally inertial space and time can be associated with a non-trivial global distribution of matter.*
- *This global distribution is necessarily fractal with $D = 2$.*

That is, according to this simple cosmology, there is a fractal $D = 2$ distribution of material, which is everywhere in dynamical equilibrium. This is entirely consistent with what is now a general concensus – at least on distance scales of $\sim 40\,\mathrm{Mpc\,h^{-1}}$.

3.2 Gravitational processes within this cosmology

The first paper (Roscoe 2002) was essentially concerned with the nature of the inertial frame. The nature of gravitational processes within the resulting equilibrium cosmology was developed in the second paper (Roscoe 2004). Basically, these were considered to arise as perturbations of the $D = 2$ equilibrium distribution of material and the specific case of a cylindrical perturbation, used as a model of an idealized spiral galaxy, was developed in detail. This development led to many successes but primary amongst these within the general MONDian context of this article are the following:

- The analysis of large samples of optical rotation curves within the context of predictions made by the theory led to the automatic recovery of the classical Tully–Fisher relations properly calibrated for both I-band and R-band photometry according to the photometry used in the sample being analyzed.
- The same analysis provided a Tully–Fisher-type relation between the characteristic radius of a spiral galaxy and its luminosity properties.
- The theory was used to model the dynamics of a sample of eight LSB galaxies provided by McGaugh and gave results that are indistinguishable from those given by the MOND algorithm. (See Fig. 13.3.)

4 Conclusions

We have discussed briefly two possible alternatives to the multi-parameter CDM paradigm – both, at the time of writing, viable.

The first, MOND, which has one free parameter (the mass-to-light ratio), has been around for about twenty years, has been shown to work extremely well in very many distinct circumstances, has a strong record of successful prediction, and

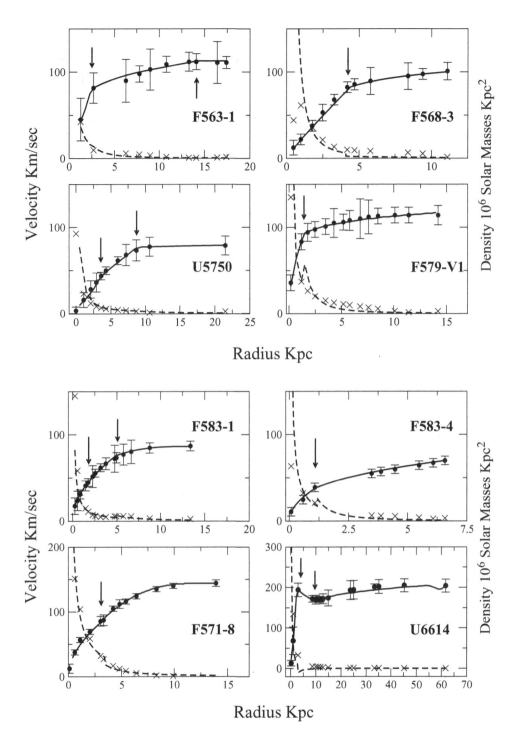

Figure 13.3 The rotation curves for a sample of eight LSBs provided by McGaugh. The rising solid lines are the computed rotation curves whilst the solid circles (with error bars) are the observations. The dashed lines are the computed mass densities ($10^6 \odot /\text{Kpc}^2$) in the disks whilst the crosses are the estimated mass densities derived from photometry.

has yet to meet unambiguous failure. It remains, however, a phenomenologically based *recipe* for calculation, which has yet to be provided with a firm theoretical foundation.

The second, developed by this author, which has two free parameters, has been around for only about four years and has been shown to work extremely well in a limited number of circumstances. It is, however, fundamentally a theoretical development based upon a particular appreciation of Mach's Principle. The comparable shared successes of MOND and this latter theory within the specific context of LSB modeling led this author to suspect that, perhaps, it is the theory underlying MOND.

References

Aaronson, M., Huchra, J., Mould, J., Tully, R. B., Fisher, J. R. *et al.*, 1982, *ApJS*, **50**, 241–262

Allen, R. J. & Shu, F. H., 1979, *ApJ*, **227**, 67–72

Bosma, A., 1978, *The Distribution and Kinematics of Neutral Hydrogen in Spiral Galaxies of Various Morphological Types*, Ph.D. Dissertation, Univ. of Groningen, The Netherlands

Casertano, S. & van Gorkom, J. H., 1991, *AJ*, **101**, 1231–1241

Corollo, C. M., Stiavelli, M., de Zeeuw, P. T., & Mack, J., 1997, *AJ*, **114**, 2366–2380

Dalcanton, J. J., Spergel, D. N., & Summers, F. J., 1997, *ApJ*, **482**, 659–676

Fish, R. A., 1964, *ApJ*, **139**, 284–305

Freeman, K. C., 1970, *ApJ*, **160**, 811–830

McGaugh, S. S., 1996, *MNRAS*, **280**, 337–354

McGaugh, S. S. & de Blok, W. J. G., 1998a, *ApJ*, **499**, 41–65

McGaugh, S. S. & de Blok, W. J. G., 1998b, *ApJ*, **499**, 66–81

McGaugh, S. S., Bothun, G. D., & Shombert, J. M., 1995b, *AJ*, **110**, 573–580

McGaugh, S. S., Schombert, J. M., Bothun, G. D., & de Blok, W. J. G., 2000, *ApJLett*, **533**, L99–L102

Milgrom, M., 1983a, *ApJ*, **270**, 365–370

Milgrom, M., 1983b, *ApJ*, **270**, 371–383

Milgrom, M., 1983c, *ApJ*, **270**, 384–389

Milgrom, M., 1984, *ApJ*, **287**, 571–576

Roberts, M. S. & Whitehurst, R. N., 1975, *ApJ*, **201**, 327–346

Roscoe D. F., 2002, *Gen. Rel. Grav.*, **34, 5**, 577–602

Roscoe, D. F., 2004, *Gen. Rel. Grav.*, **36, 1**, 3–45

Rubin, V. C., Ford, W. K., & Thonnard, N., 1980, *ApJ*, **238**, 471–487

Sanders, R. H., 1986, *MNRAS*, **223**, 539–555

Sanders, R. H. & McGaugh, S. S., 2002, *ARAA*, **40**, 263–317

Sellwood, J. A. & Kosowsky, A., 2001, In *Gas and Galaxy Evolution*, ASP Conf. Series, eds. J. E. Hibbard, M. P. Rupen, J. H. van Gorkom, pp 311–318. San Francisco: Astronomical Society of the Pacific

Shostak, G. S., 1973, *Astron. Astrophys.*, **24**, 411–419

Tully, R. B. & Fisher, J. R., 1977, *A&A*, **54**, 661–673

van Albada, T. S. & Sancisi, R., 1986, *Phil. Trans. Roy. Soc. London A.*, **320**, 447

14

Anthropic principle in cosmology

Brandon Carter

LuTh, Observatoire de Paris-Meudon, France

Abstract

A brief explanation of the meaning of the anthropic principle – as a prescription for the attribution of a priori probability weighting – is illustrated by various cosmological and local applications, in which the relevant conclusions are contrasted with those that could be obtained from (less plausible) alternative prescriptions such as the vaguer and less restrictive ubiquity principle, or the more sterile and restrictive autocentric principle.

Introduction

Having been asked to contribute a discussion of the anthropic principle for a colloquium on cosmology, I would start by recalling that although its original formulation [1] was motivated by a problem of cosmology (Dirac's) and although many of its most interesting subsequent applications (such as the recent evaluation [2] of the dark energy density in the Universe) have also been concerned with large scale global effects, the principle for which I introduced the term "anthropic" is not intrinsically cosmological, but just as relevant on small, local scales as at a global level. In retrospect I am not sure that my choice of terminology was the most appropriate, but as it has now been widely adopted [3] it is too late to change. Indeed the term "anthropic principle" has become so popular that it has been borrowed to describe ideas (e.g., that the Universe was teleologically designed for our kind of life, which is what I would call a "finality principle") that are quite different from, and even contradictory with, what I intended. This presentation will not attempt to deal with the confusion that has arisen from such dissident interpretations, but will be concerned only with developments of my originally intended meaning, which I shall attempt to explain in the next section.

173

Meaning of the anthropic principle

Whenever one wishes to draw general conclusions from observations restricted to a small sample it is essential to know whether the sample should be considered to be biased, and if so how. The anthropic principle provides guidelines for taking account of the kind of bias that arises from the observer's own particular situation in the world.

Although frequently relevant to purely local applications, the anthropic principle was originally formulated in a cosmological context as a reasonable compromise to two successively fashionable extremes. The first of these was what might be described as the autocentric principle, which underlay the pre-Copernican dogma to the effect that as terrestrial observers we occupy a privileged position at the center of the Universe. The opposite extreme was the more recent precept describable as the cosmological ubiquity principle, but commonly referred to just as the cosmological principle, which would have it that the Universe is much the same everywhere, having no privileged center, and that our own neighborhood can be considered as a typical random sample.

To put it more formally, in conventional Bayesian terminology, the a-priori probability distribution for our own situation was supposed, according to the autocentric principle, to have been restricted to the region where we actually find ourselves, whereas according to the ubiquity principle it was supposed to have been uniformly extended over the whole of space-time. Thus according to the autocentric principle we could infer nothing at all about the rest of the Universe from our local observations, whereas according to the ubiquity principle we could immediately infer that the rest of the Universe was fairly represented by what we observe here and now.

As a reasonable compromise between these unsatisfactory over-simplistic extremes, the anthropic principle would have it that – within the context of whatever theoretical model may be under consideration – the a-priori probability distribution for our own situation should be prescribed by an anthropic weighting, meaning that it should be uniformly distributed, not over space-time (as the ubiquity principle would require), but over all observers sufficiently comparable to ourselves to be qualifiable as anthropic.

Of course if the qualification "anthropic" were interpreted so narrowly as to include only members of our own human species, then the cosmological implications of the anthropic principle would reduce to those of the scientifically sterile autocentric principle, but it is intended that the term "anthropic" should also include extraterrestrial beings with comparable intellectual capabilities. Thus, unlike the autocentric principle, but like the ubiquity principle, the anthropic principle has non-trivial implications that can be subjected to empirical verification. The proto-type example was provided by the famous debate [4] between Dirac and Dicke about whether the strength of gravitation should decrease in proportion to the expansion of the Universe. Subsequent work has shown rather conclusively that Dirac's prediction (that it would), which was implicitly based on the cosmological ubiquity principle, must be rejected in favour of Dicke's prediction (that it would not), which was implicitly based on the anthropic principle. (This debate illustrates a common source of misunderstanding in this area, which is that relevant but questionable principles tend to be taken for granted tacitly, and even subconsciously, rather than being made explicit.)

If it were necessary to be more precise, one would need some kind of *microanthropic* principle specifying the notion of anthropic weighting in greater detail, dealing with questions such as whether it should be proportional to the longevity and erudition of the individuals under consideration. (For example, should someone like Dirac or Dicke qualify for a higher weighting than a child who dies in infancy before even learning to count?) I have recently shown [5] how this issue provides insights that are useful for the fundamental problem of the interpretation of quantum theory.

The strong anthropic principle

For the crude qualitative applications of the anthropic principle that have been discussed so far in the scientific literature, the fine details dealt with by the microanthropic principle [5] are in practice unimportant.

There is, however, a refinement of a rather different kind that plays a significant role in the published literature. This is the distinction between what are known as the "strong" and "weak" versions of the anthropic principle. In the ordinary, widely accepted, "weak" version the relevant (anthropically weighted) a-priori probability is supposed to concern only a particular given model of the Universe, or a part thereof, with which one may be concerned. In the more controversial "strong" version the relevant anthropic probability distribution is supposed to be extended over an ensemble of cosmological models that are set up with a range of different values of what, in a particular model, are usually postulated to be fundamental constants (such as the well-known example of the fine structure constant). The observed values of such constants might be thereby explicable if it could be shown that other values were unfavorable to the existence of anthropic observers. However if (as many theoreticians hope) the values of all such constants should turn out to be mathematically derivable from some fundamental physical theory, then the "strong" version of the anthropic principle would not be needed.

A prototype example of the application of this "strong" kind of anthropic reasoning was provided by Fred Hoyle's observation [6] that the triple alpha process, which is necessary for the formation (from primordial hydrogen and helium) of the medium and heavy elements of which we are made, is extremely sensitive to the values of the coupling constants governing the relevant thermonuclear reactions in large main sequence stars. This contrasts with the case of the biochemical processes (depending notably on the special properties of water) to which such considerations do not apply, despite the fact that (as discussed by Barrow and Tipler [7]) they are also indispensible for our kind of life. The relevant biochemical properties are not sensitive to the values of any physical coupling parameters but are mathematically determined by the quantum mechanical consequences of the special properties of the three-dimensional rotation group.

Although it does not affect the chemistry of the light and medium weight elements that play the dominant role in ordinary biochemistry, the particular value (approximately 1/137) of the electric coupling parameter that is (appropriately) known as the "fine structure" constant is more significant for the – less biologically relevant – details of heavy element chemistry. Of potentially greater "strong" anthropic relevance, however, is the effect [1] of the "fine structure" constant on the convective instabilities that are probably important for the creation of planets during main sequence star formation.

A particularly topical application [2, 8] of the "strong" anthropic principle concerns the recently estimated value of Einstein's cosmological repulsion constant on the supposition that it is identifiable with what is commonly referred to as the "dark energy density" of the Universe. If this parameter had been much larger (as might have been naively expected from fundamental physical considerations) then the Universe would have already been inflated to such a low density at such an early stage in its life after the big bang that the galactic and stellar structures needed for our life systems would never have been able to condense out at all.

Although far from tautological, but of considerable scientific interest from the point of view of explaining the environment in which we find ourselves, the foregoing examples do not actually provide direct predictions of facts that are not already well established. However, the next section will describe examples in which the anthropic principle provides genuine predictions in the form of conclusions that remain unconfirmed and even controversial.

Anthropic prediction

Although oversimplified expressions of the anthropic principle (such as the version asserting that life exists only where it can survive) reduce to mere tautology, the more complete formulation (prescribing an a-priori probability distribution) can provide non-trivial predictions that may be controversial, and that are subject to rational contestation since they are different from what would be obtained from alternative prescriptions for a-priori probability, such as the ubiquity principle that would attribute a-priori (but of course not a-posteriori) probability even to uninhabited situations.

The example that seems to me most important was provided by the prediction [9] that the occurrence of anthropic observers would be rare, even on environmentally favorable planets such as ours. This prediction was based on the observation that our evolutionary development on Earth has taken a substantial fraction of the time available before our Sun reaches the end of its main sequence (hydrogen burning) life. This would be inexplicable on the basis of the ubiquity principle, which would postulate that the case of our planet was typical and hence that life like ours should be common. On the basis of the anthropic principle it would also be inexplicable if one supposes that biological evolution can proceed easily on time scales short compared with those of stellar evolution, but it is just what would be expected if the biological evolution of life like ours depends on chance events with characteristic time scales long compared with those of stellar evolution.

The (as yet unrefuted) implication that I drew from this (more than twenty years ago) was that the search for extraterrestrial civilizations was unlikely to achieve easy success. I have found, however, that such conclusions tend to be unpopular in many quarters, presumably because they involve limitations on the extent and more particularly the duration of civilizations such as ours, which (in lieu of personal immortality) many people would prefer to think of as everlasting. In the words of Dirac (when refusing to accept Dicke's effectively anthropic reasoning [4]) the assumption to be preferred is "the one that allows the possibility of eternal life."

One of the most remarkable attempts to show that – despite the inexorable action [10] of the entropy principle commonly known as the Second Law of thermodynamics – life could after all continue to exist in the arbitrarily distant future, has been made by Freeman Dyson [11], whose recent intervention in a related debate [12] provides another striking example of the kind of misunderstanding the anthropic principle was meant to help avoid. However, the issue on this occasion is not the very-long-term future of life in the Universe, but the more immediate question of the future of our own terrestrial civilization in the next few centuries. Apparently under the influence of wishful thinking reminiscent of Dirac's, Dyson has strongly objected to a thesis developed particularly by Leslie [13] (and from a slightly different point of view by Gott [14]) of which a conveniently succinct discussion with a comprehensive review of the relevant literature was provided by Demaret and Lambert [15]. The rather obvious conclusion in question is that the anthropic principle's attribution of comparable a-priori weighting to comparable individuals within our own civilization makes it unlikely that we are untypical in the sense of having been born at an exceptionally early stage in its history, and hence unlikely that our civilization will contain a much larger number of people born in the future.

The foregoing reasoning implies that our numbers will either be cut off fairly soon by some (presumably [9] man-made, e.g., ecological) catastrophe (the "dooms-day" scenario [13]) or else (more "optimistically") will be subject to a gradual (controlled?) decline that must start even sooner but that could be relatively pro-longed. Despite the fact that such conclusions can be and have been drawn inde-pendently (without recourse to anthropic reasoning) from other considerations of an economic or environmental nature, Dyson persists [12] in denying their valid-ity, thereby implicitly repudiating the anthropic weighting principle on which they are based. Dyson's position seems to be based on what might be called the "auto-centric principle" (the extreme opposite to the "ubiquity principle") as referred to above, whereby one attributes a-priori probablity only to one's actual position in the Universe. A supposition of this commonly (but usually subconsciously) adopted kind makes it legitimate for Dyson to rule out the use of the Bayes rule as something that is redundant (albeit not strictly invalid) because, according to this autocentric principle, no a-priori probability measure is attributable to anything inconsistent with what has already been observed. However, (quite apart from its failure to face the ecological considerations leading to the same conclusions) Dyson's wishful thinking in this context seems even less intellectually defensible than Dirac's ubiq-uitism, because the autocentric principle effectively violates Ockham's razor by its solipsistic introduction of an artificial distinction between "oneself" and other manifestly comparable observers.

References

[1] B. Carter, Large number coincidences and the anthropic principle in cosmology, in *Confrontations of Cosmological Theories with Observational Data* (I. A. U. Symposium 63) ed. M. Longair, 291–298 (Reidel, Dordrecht, 1974).

[2] J. Garriga, A. Linde, A. Vilenkin, Dark energy equation of state and anthropic selection, *Phys. Rev.*, **D69**, (2004), 063521. [hep-th/0310034]

[3] N. Bostrom, *Anthropic Bias: Observation Selection Effects in Science and Philosophy* (Routlege, New York, 2002).

[4] R. H. Dicke, Dirac's cosmology and Mach's principle, *Nature*, **192**, (1961), 440–441.

[5] B. Carter, Anthropic interpretation of quantum theory, contrib. to 2003 Peyresq Physics Meeting "The Early Universe". [hep-th/0403008]

[6] F. Hoyle, D. N. F. Dunbar, W. A. Wenzel, W. Whaling, A state in C^{12} predicted from astrophysical evidence, *Phys. Rev.*, **92**, (1953), 1095.

[7] J. F. Barrow, F. J. Tipler, *The Anthropic Cosmological Principle* (Clarendon Press, Oxford, 1986).

[8] L. Pogosian, A. Vilenkin, M. Tegmark, Anthropic predictions for vacuum energy and neutrino masses, *J. Cosm. Astropart. Phys.*, **0407**, (2004), 005. [astro-ph/0404497]

[9] B. Carter, The anthropic principle and its implications for biological evolution, *Phil. Trans. Roy. Soc.*, **A310**, (1983), 347–363.

[10] J. Islam, Possible ultimate fate of the universe, *Q. J. Roy. Astr. Soc.*, **18**, (1977), 3–8.

[11] F. J. Dyson, Time without end: physics and biology in an open system, *Rev. Mod. Phys.*, **51**, (1979), 447–460.

[12] F. J. Dyson, Reality bites, *Nature*, **380**, (1996), 296.

[13] J. Leslie, Time and the anthropic principle, *Mind*, **101**, (1992), 521–540.

[14] J. R. Gott III, Implications of the Copernican principle for our future prospects, *Nature*, **363**, (1993), 315–319.

[15] J. Demaret, D. Lambert, *Le Principe Anthropique* (Armand Colin, Paris, 1994).

Discussion

Comment:

A. BLANCHARD :

I was wondering how much your conclusion is not entirely built on the assumptions that you start from (i.e., you get the parameters that are more or less those that correspond to the values that triggered the assumptions used). For instance, if you accept that non-linear objects might form when the Universe is 10 seconds, life will still be possible with very different cosmological parameters.

Comment :

F. SANCHEZ

I have shown you a project paper where there is a number of very precise (10^{-2} to 10^{-4}) correlations, involving large numbers, based on an elementary Dimensional Analysis. Since the cosmologic application of your biologic anthropic principle is based on rough coincidences (one order magnitude) I refute such an interpretation.

Part VI

Evidence for anomalous redshifts

15

Anomalous redshifts

H. Arp

*Max-Planck-Institut für Astrophysik, Karl Schwarzschild-Str.1, Postfach 1317, D-85741
Garching, Germany*

Abstract

The time has come to stop calling measured redshifts of extragalactic objects
"anomalous" or "discordant." Observational evidence over 38 years has made it
clear that objects at the same distance from the observer can have strongly differing
redshifts. Rigorous solutions of the basic mass, energy, and momentum equations
show that redshift is primarily a function of the age of the matter constituting
a galaxy. Reluctance to accept these results is blocking meaningful advance in
physics and cosmology.

1 Introduction

Starting in 1966 evidence began to accumulate that high redshift quasars were
physically associated with low redshift, relatively nearby galaxies. Of course the
existence of even one redshift not caused primarily by recession velocity would
negate the fundamental assumption on which all big-bang cosmology depends. In
the ensuing 38 years a majority of extragalactic astronomers have built a complex
and massively publicized edifice on the assumption that redshifts are an identical
measure of distance. During these same decades a minority of astronomers have
struggled to observe and report the increasingly powerful evidence that contradicts
that crucial assumption.

In the present review we show only samples of this contradictory observational
evidence taken from a body of evidence that is now too large for even book-sized
discussions. Once the empirical rules of association are laid down the pictures and
diagrams communicate at a glance more eloquently than text. As a result we will
communicate here the main thread of the argument in pictorial form.

Some criticisms that are used against this evidence can be refuted in advance:

1. Argument

Quasars are the nuclei of host galaxies that have the same redshift. Since the host galaxies
   must be at their redshift distance, quasars must also be at their redshift distance.

Answer:

When the material that makes up the quasar expands into a resolvable galaxy, all the
   material has the same intrinsic redshift. Evolution of quasars into galaxies is supported
   by much observational evidence that the redshift of essentially the whole of younger
   companion galaxies is intrinsic.

2. Argument

Alignments of quasars and other high redshift objects across galaxies are not meaningful
   because the probability estimates are a posteriori.

Answer:

The first alignment is a posteriori. Each subsequent alignment is confirmation of an
   a-priori prediction. Moreover the improbabilities compound as each new case is discovered. The stated characteristics are closeness of bright objects, their alignment,
   their centering, and the similarity of the aligned objects. These are all properties
   expected of objects ejected from active galaxies.

The empirical evidence for this picture is statisically so strong that it is embar-
rassing to mention numbers. In addition there is much direct evidence of high
redshift objects linked by optical, X-ray, and radio filaments to active galaxies of
much lower redshift. Some samples will be shown here and some discussion of
the observational consequences of ejection of low particle mass matter will be
given.

The overridingly important message of this presentation is, however, that the conventional assumption about extragalactic redshifts is drastically incorrect. Almost
every current conclusion about cosmology, origins, and physical processes in the
Universe needs to be reassessed and replaced.

2 A quasar 8 arcsec from the nucleus of the Seyfert NGC 7319

We start with the most recent discovery of a high resolution Chandra X-ray source
very close to the nucleus of the very active Seyfert Galaxy in Stephan's Quintet.
The Seyfert is ejecting radio plasma, X-rays, and emission line gas. Figure 15.1
shows an optical filament coming out of the nucleus and bending toward the quasar,
stopping about 2 arcsec from it. The galaxy has a redshift of $z = 0.0225$ and the
quasar has been observed with a Keck 10 meter telescope to have a redshift of
$z = 2.11$ (see for details P. Galiani, E. M. Burbidge, H. Arp, G. R. Burbidge, V.
Junkkarinen, and S. Zibetti, *ApJ*, **620**, 88, 2005).

There are four arguments one can make that these two objects of vastly different
redshift are physically interacting:

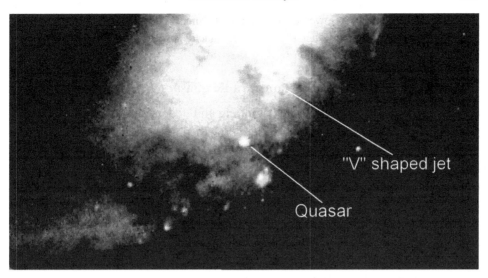

Figure 15.1 HST image of the central region of NGC 7319 showing optical extension to the ULX/quasar 8 arcsec south of the nucleus. The redshift of the galaxy
is $z = 0.022$ and the quasar $z = 2.11$.

1. The significance of X-ray sources of this strength clustering within $10'$–$25'$ distance on
   the sky around bright Seyferts is of the order of 7.5 sigma (Radecke 1997, Arp 1997).
   But the distance of the NGC 7319 source from its nucleus is only $8''$, which is an area
   3.3×10^{-5} smaller. Therefore the probability of an unrelated X-ray source falling this
   close to the Seyfert nucleus is negligible. Their proximity is the reason why such sources,
   called ultra-luminous X-ray sources (ULXs), have been routinely regarded as black hole
   binaries belonging to the galaxy. (But see Arp *et al.*, 2004 where 24 out of 24 have been
   measured to be high redshift quasars.)
2. The Seyfert inner regions are so dusty and obscured that it is extremely unlikely that a
   backround object could show through. The quasar shows no indication of being exces-
   sively reddened.
3. The gas in the end of the galaxy in which the quasar is situated shows strong forbidden
   emission lines. In fact the low density forbidden line $[OII]$ is so strong and extended
   that it must form some sort of a low density halo around that end of the galaxy. The only
   source of ionization would be the quasar pumping ionizing photons into this feature from
   its redshifted Lyman alpha and blueward continuum. In fact Aoki *et al.* (1996) *without
   knowledge of the* $[OII]$ *line or the quasar* calculated at least an order of magnitude
   deficit of ionizing photons just for the rest of the emission lines in this end of the NGC
   7319 inner regions.
4. The X-ray material, the radio material, and the excited gas are all rushing out of this
   end of the Seyfert nucleus. The compact synchrotron source, the quasar, seems also to
   have come out with this ejection of material. The observations are concordant with all
   previous evidence for ejection of quasars, except now the high resolution X-ray and

optical telescopes have allowed us to look into the inner regions where the high redshift quasar is just starting its voyage out into extragalactic space.

3 Ejection of radio quasars

It is now possible to go back to the beginning and understand more of the various proofs of an ejection origin of quasars and how they fit together in a useful model. It was long ago accepted that radio emitting material is ejected, usually paired in opposite ejections, from the centers of active galaxies. The major additional property of the subsequent, so-called anomalous, observations is that some of this material can have much higher redshift than the central galaxy.

The strong pairs of radio sources like the famous 3C273 and 3C274 across the brightest galaxy in the Virgo Cluster (Arp 1966, 1967) first indicated that a quasar could be at the same distance as a nearby, low redshift galaxy rather than at the much larger distance indicated by its redshift. But when the all-sky surveys at Parks and Cambridge were completed and investigated spectroscopically it became possible to work with complete samples of radio quasars.

Figure 15.2 shows a pair of 3C radio quasars across the disturbed pair of galaxies NGC 470/474 (Arp, *Atlas of Peculiar Galaxies No. 227*). Quasars this radio bright are very rare (a total of 50 over the northern hemisphere). This yields a frequency of only one per 320 sq. deg. and a chance of only 5×10^{-6} of finding both so close to an arbitrary point in the sky. We then calculate the chance that they are also accidentally aligned within a degree or so, that they are equally spaced across the centroid within about 10%, and that their redshifts are within 0.09 of each other out of a range of about 2. All this combines to a probability of about 3×10^{-8} that this can be an accidental association (about three chances in a hundred million). Note that this is not an a-posteriori probability because for 38 years many examples of paired quasars across active galaxies have been found with just these characteristics. In fact it is the confirmation of a predicted configuration at a significance level that should be considered conclusive.

An almost identical association is shown in Fig. 15.3. There the central object is a bright starburst galaxy with a blue jet arm extending out to the WNW. Its redshift is $z = 0.009$, very much like the $z = .008$ of the galaxies in the preceding Fig. 15.2. Also as in the preceding example, two 3C radio quasars of very similar redshift ($z = 0.22$ and 0.29) are paired at only slightly greater distances across the active galaxy. To have two such associations at this probability level is extraordinarily compelling. And, of course, very similar pairings are found in the southern hemisphere where the bright radio sources are Parks rather than 3C Cambridge Survey Sources (see *Catalog of Discordant Redshift Associations*; Arp 2003).

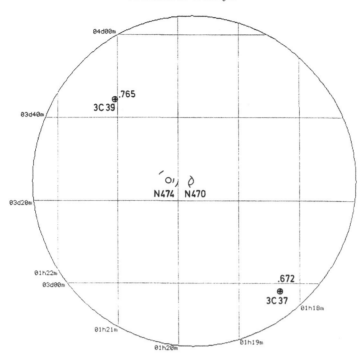

Figure 15.2 NGC 470/474 (Arp 227). Two disturbed galaxies with a pair of strong, 3C radio quasars ($z = .765$ and $.672$) paired across them. Accidental probability of this alignment is 3×10^{-8}.

As a culmination to this sample of galaxy/radio quasar associations we show the closest known in Fig. 15.4 – a 3C radio source which *had two redshifts*. In the paper reporting this (Tran *et al.* 1998), the abstract ended with the statement "Our data reveal a chance alignment of 3C 343.1 with a foreground galaxy, which dominates the observed optical flux from the system." It was a simple matter, however, to look up the high resolution radio map (Fanti *et al.* 1985) and find the two objects linked together by a radio bridge as shown here in Fig. 15.4. We now calculate some probabilities of this being a chance alignment and show how the configuration follows the rules of many previous physical associations.

A circle of 0.25 arcsec radius subtends an area of 1.5×10^{-8} sq. deg. on the sky. In the now essentially completely identified 3C Catalog there are about 50 radio quasars. Assuming 23 000 sq. deg. to Dec. $= -5$ deg., we compute 2.2×10^{-3} such quasars per sq. deg. This gives a probability of 3×10^{-11} of accidentally finding the $z = 0.750$ quasar within 0.25 arcsec of the $z = 0.344$ galaxy.

However, even if we do not consider the radio material linking them a conclusive physical bridge, we must still estimate the accidental possibility that the radio tail

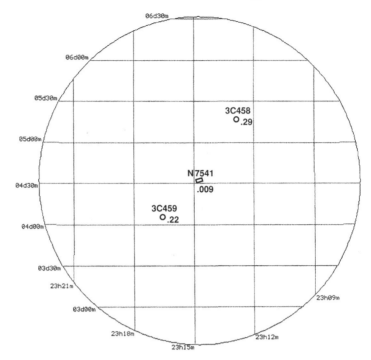

Figure 15.3 NGC 7541, a bright starburst galaxy with a blue jet/arm and $z = 0.009$ falls between two strong, 3C radio sources, which are now identified as quasars with $z = 0.29$ and 0.22.

from the galaxy points within a few degrees to the quasar and similarly from the quasar back to the galaxy. This would give a further improbability of $(\pm 2/90)^2 = 5 \times 10^{-4}$. *The combined probability of this configuration being chance is of the order of* 10^{-14}.

4 Ejection of X-ray quasars

One of the reasons we know radio sources are ejected is that we observe radio emitting gas moving outward in jets that terminate on extended clouds of radio emission. Since the advent of X-ray astronomy we can also observe narrower, higher density X-ray jets emerging from active galaxies, some as narrow cores to the radio ejections. In the case of pairs and lines of X-ray sources ejected from active galaxies, however, almost every point X-ray source can be established as a high redshift quasar. Why do many radio sources appear as blank fields with no optical object?

In the past Arp has suggested that radio/X-ray quasars on their way out of the inner regions of galaxies or through the intergalactic medium are stripped of their

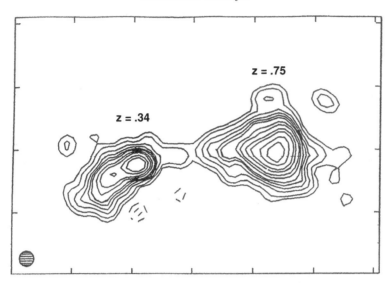

Figure 15.4 Radio map at 1.6 Ghz of 3C343.1 by Fanti *et al.* 1985. *Separation of sources is only 0.25 arcsec.* Note the opposite ejections from the radio galaxy, the western of which leads directly into the quasar. The compression of the radio contours on the west side of the quasar attests to its motion directly away from the galaxy.

outer layer of lower density, radio emitting gas (Arp 2001a,b; 2003). This would account for numbers of radio sources with no optical identification but almost all of the denser X-ray sources having optical identifications. Because the optical objects are invariably quasars it is possible to study the properites of ejected quasars by studying X-ray sources associated with active galaxies.

Figure 15.5 shows all the bright X-ray sources in the field of the very active Seyfert galaxy NGC 3516. They are distributed in a line through the nucleus of the galaxy and every one of them turns out to be a high redshift object. They display the less disturbed properties of ejected quasars in that the redshifts are highest closest to the galaxy and progressively drop as they move outward. NGC 3516 and another Seyfert with a line of quasars display exactly this property (Arp 1999). These two Seyferts also both display the property that the quasars follow the line of the projected minor axes of the galaxy. This is important because along the minor axis ejecta can come out with the least interaction with the material of the galaxy.

It is also important to note that in these two Seyferts 10 out of their 11 quasars fall close to the periodic values of redshift (Karlsson formula for preferred, peak redshift values). Although reasons for the periodicity have not been agreed (perhaps Machian contacts at velocity c with a fractal universe) it is clear that these

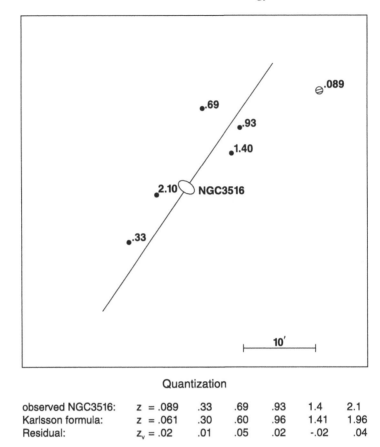

Quantization

observed NGC3516:	z = .089	.33	.69	.93	1.4	2.1
Karlsson formula:	z = .061	.30	.60	.96	1.41	1.96
Residual:	z_v = .02	.01	.05	.02	-.02	.04

Figure 15.5 The Rosetta Stone. The brightest X-ray sources in the field are aligned along the minor axis in descending order of quantized redshift. Very active Seyfert has $z = 0.009$.

observations rule out velocity caused redshifts. Figure 15.5 seems to be the Rosetta Stone that unlocks the birth and evolution of galaxies.

Figure 15.6 shows the averaged properties of associations studied up to about 1998. It schematically outlines the ejection characteristics and the relation of proto quasars to mature galaxies. It explains the enigmatic result of Erik Holmberg in 1969 that companion galaxies were characteristically spread along the minor axis of edge-on larger galaxies. The excess redshifts of compact companion galaxies in groups (Collin-Souffrin, Pecker, and Tovmassianis 1974) is also vindicated. The redshifts are quantized and shown moving rapidly between discrete values. The diagram in Fig. 15.6 represents an empirical theory summarizing the evolution of extragalactic objects.

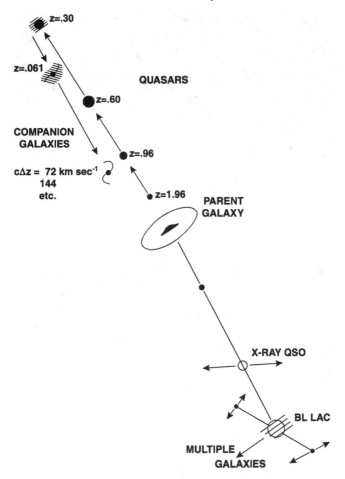

Figure 15.6 Extracting the characteristics of all associations of quasar and companion galaxies up to about 1998. Pictured is an empirical theory of galaxy birth and evolution from quasars.

5 Intrinsic redshifts and evolution of galaxies

Another Seyfert galaxy that is an active strong X-ray source is NGC 7603. It is conspicuous for the large companion galaxy attached to a curved filament or arm, which extends from the disturbed body of the galaxy. The Seyfert has a redshift of $z = 8000$ km s^{-1} and the companion $16\,000$ km s^{-1}. The obvious interpretation is that the compact proto companion was ejected some time ago and encountered interaction with the material of the galaxy (the interaction tail in Fig. 15.7). Instead of evolving to its present state at a greater distance from the galaxy it evolved to only slightly younger age close by and has now an intrinsic redshift relative to its parent of only 8000 km s^{-1}.

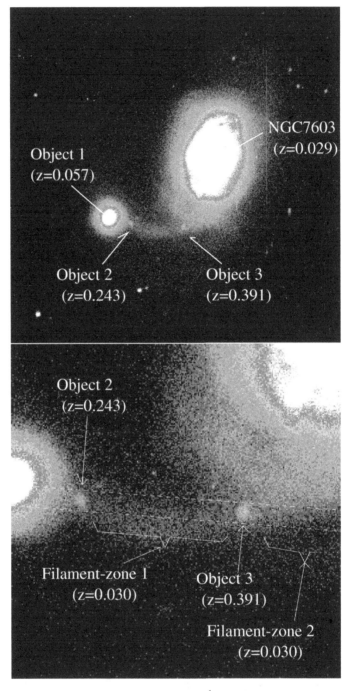

Figure 15.7 NGC 7603 at $cz = 8000$ km s^{-1} attached to companion at $cz = 16\,000$ km s^{-1}. Compact objects in the arm from the parent Seyfert turn out to be high redshift quasar-like objects (López-Corredoira and Gutiérrez 2002).

Looking at the pictures immediately raised the question of whether two faint compact objects visible in this ejection track represent even younger material with even higher intrinsic redshift. It took 30 years to answer this obvious question but finally two courageous young Spanish astronomers took the spectra of these objects and found them to be quasars of $z = 0.24$ and 0.39. There was some quibbling by conventional astronomers about calling them quasars because they were slightly non-stellar and the emission lines were somewhat thin. But as we see in the following section we would expect just this result if quasars interacted with the galaxy medium and were ablated or slightly deformed and gas emitting regions were slightly diffused.

It is significant that the original discovery of the attached companion to NGC 7603 caused Fred Hoyle to propose a fundamentally new approach to physics where particle masses grew as t^2 and that enabled the Hubble law to be derived without expansion of the Universe. The prestigious Russell Lecture of 1972 in which he set forth this paradigm-breaking advance was never accepted for publication in the *Astrophysical Journal*. (See my note in *APSS*, **285**, 451, 457.) We discuss now in the closing paragraphs how this theory has grown unnoticed to give a rational understanding of the enormous amount of currently ignored data.

6 The physical cause of age related redshifts

The key to acceptance of the observations, however, seems to require a theory that leads to an understanding of what gives the quasars their intrinsic redshifts. We invoke the variable mass hypothesis as developed by Narlikar (1977) and Narlikar and Arp (1993). In a more general solution of the general relativistic field equations the particle masses of new matter start out at or near zero mass and grow with time. Because the electrons making orbital transitions in radiating atoms are initially small, the emitted photons are initially redshifted and decrease their intrinsic redshift with time. The quasars are then viewed as being composed of young matter, which evolves toward normal matter and normal galaxies with time.

In the initially ejected proto quasars the particles grow in mass and slow down in order to conserve momentum so the particles cool and increasingly gravitate toward a young galaxy (no dark matter needed). The initial plasmoid, however, has low mass ions, which have large cross sections and are initially constrained by magnetic fields (Arp 1963). In order to have the time to evolve intrinsic redshifts into the range of older galaxies like our own they must be slowed down or stopped by the passage through the internal regions of the parent galaxies or meeting clouds in the medium exterior to the galaxy. In the proto-quasar phase they should be fragile and observations suggest that some quasars are split into twos or threes (Arp and Russell 2001, p. 548, Arp 1997, 1999). In practice it is suggested that sometimes they can run into a medium of cloudlets and be divided into many small proto galaxies i.e., a cluster of proto galaxies on its way to evolving into a cluster of galaxies.

It is worth noting that active galaxies, quasars, and clusters of galaxies are the three principal kinds of extragalactic X-ray sources that exist. In the above picture each are subunits of the former. The processes in galaxy nuclei that give rise to the quasars furnish the energy to fission or explode some quasars into smaller pieces, which then evolve into galaxy clusters, particularly in interaction with a galaxy/extragalactic environment.

7 Summary

The greatest chance for progress in cosmology and physics is now to recognize that redshifts can not be "anomalous" but instead mark the change with time of the evolutionary age of the objects that make up the Universe.

References

Aoki, K., Ohtani, H., Yoshida, M., and Kosugi, G., 1996, *AJ*, **111**, 140

Arp, H., 1963, *Scientific American*, **28**, p. 71

Arp, H., 1966, *Science*, **151**, 1214

Arp, H., 1967, *ApJ*, **148**, 32

Arp, H., 1997, *A&A*, **327**, 479

Arp, H., 1999, *A&A*, **341**, L5

Arp, H., 2001a, *ApJ*, 549, 780

Arp, H., 2001b, *Recontres de Moriond, http://www-dapnia.cea.fr/Conferences/ Morionastro2001/index.html*

Arp, H., 2003, *Catalog of Discordant Redshift Associations*, Apeiron, Montreal

Arp, H. and Russell D., 2001, *ApJ*, **549**, 802

Arp, H., Gutierrez, C., Lotez, M., 2004, *A & A* **418**, 877

Collin-Souffrin, S., Pecker, J.-C., and Tovmassian, H., 1974, *A&A*, **30**, 351

Fanti, C., Fanti, R., and Parma, P., 1985, *A&A*, **143**, 292

Hawkins, E., Maddox, S., and Merrifield, M., 2002, *MNRAS*, **336**, 13L

Lòpez-Corredoira M. and Gutièrrez, C., 2002, *A&A*, **390**, 15L

Narlikar, J.V., 1977, *Ann Phys-New York*, **107**, 325

Narlikar, J. and Arp, H., 1993, *ApJ*, **405**, 51

Narlikar, J.V. and Das, P.K., 1980, *ApJ*, **240**, 401

Radecke, H.-D., 1997, *A&A*, **319**, 18

Tran, H., Cohen, M., Ogle, P., *et al.*, 1998, *ApJ*, **500**, 660

Discussion

Q : A. BLANCHARD :

A few years ago, Tom Broadhurst did claim to find a periodicity in a sample of about 1000 galaxies. This periodicity first seen on a sample of a few hundred galaxies seems to be stronger in the larger samples. However, it seems to have entirely disappeared now with larger samples with more than 10 000 galaxies. So I claim that it is likely that when the number of objects increases by a factor of 10, the signal will just vanish!

A : H. A. :

The periodicities in redshifts of galaxies has been reviewed by Napier in this conference. You seem to be referring to Hawkins, Maddox, and Merrifield (*MNRAS* 2002) who analyzed a sample of over 22 000 faint quasars and reported for the first time in 37 years that there was no periodicity in their redshifts. Napier and G. Burbidge pointed out in a subsequent issue of *MNRAS* (**342**, 60, 2003) that they had not tested against the active parent galaxies into whose redshift frame these faint quasars had to be transformed. But the press release that the Hawkins *et al.* authors had issued was accepted by most of the people in the field as discounting all previous periodicity evidence.

In 2003, an extensive analysis of this same faint sample of quasars by Arp, Roscoe, and Fulton demonstrated that the standard periodicities were clearly present in physical associations of quasars in this survey. This paper was rejected by the Monthly Notices. This perhaps enables us to understand how it comes about that concordance with big bang assumptions continues to be accepted in the face of massive and growing observational contradictions.

Q : J. SULENTIC :

The X-ray quasar discovered near the nucleus of the Seyfert 2 galaxy NGC 7319 may represent an important link between the high redshift quasars apparently registered around low redshift galaxies. If Arp is correct and the low z galaxies eject these quasars, then we might expect (hope?) to observe some in close proximity to the galactic nucleus. At least two other candidates are known, involving (as in the cases of NGC 7319) a hard X-ray source very near (few arcsec) a Seyfert nucleus.

A : H. A. :

If these objects are optically identified, it would be very important to get their spectra.

Q : J. V. NARLIKAR :

Arp's findings on intrinsic redshifts can be understood in terms of the variable mass hypothesis proposed by me and discussed in detail by Narlikar and Das (1980). In this the ejected quasar/galaxy has new matter whose mass (per particle) increases with age, and so intrinsic redshift declines with age. So companion galaxies are the "aged" version of quasars ejected from parent galaxies.

However, I have not yet managed to quantize that picture to arrive at an explanation of periodicity.

A : H. A. :

In the Hoyle/Narlikar approach, the masses grow by exchanging (gravitons, machions) in a volume of space, which increases in radius at the speed of c. If the Universe had hierarchical distribution of distances, one could obtain predominantly discrete redshift values that would evolve in jumps.

16

Redshifts of galaxies and QSOs: The problem of redshift periodicities

Geoffrey Burbidge

University of California, San Diego, La Jolla, CA, USA

1 The redshifts of normal galaxies

For more than 70 years observational evidence has been steadily accumulated that shows that the original observations of Hubble, which led directly to the view that the Universe is expanding, apply to normal galaxies made up of stars. Hubble's original redshift-apparent magnitude relation of 1929 was steadily extended to fainter galaxies, so that by the 1950s it covered a range from about $cz \simeq 1000$ km s^{-1} to values of z close to 0.2 (Humason, Mayall, and Sandage 1956). By about 1960, following the discovery of the radio galaxies, Minkowski (1960) had reached a redshift record with the galaxy associated with 3C 295, which has $z = 0.46$. In the 1960s it was very difficult to go beyond that. The limits were set by sizes of the telescopes, the efficiency of the detectors, the faintness of the galaxies, and ways of finding suitable distant clusters. These barriers were all eventually overcome, and for galaxies we can now confidently extend the Hubble law out to galaxies with $z \simeq 3$.

However, while this redshift-apparent magnitude relation taken in the large is apparently a smooth function of z, Tifft showed in the early 1960s, first by studying the redshifts in the Coma cluster of galaxies, that the differential redshifts Δz among the different galaxies in a cluster appeared to be quantized, so that the redshift differences are of the form $n\Delta z$, with $c\Delta z, \simeq 72$ km s^{-1}, and n is an integer.

While the reality of this effect was doubted on all sides, it was confirmed by Weedman, and also the same effect was found by others in pairs of galaxies, and in the redshift differences between satellite galaxies and the central galaxies in small groups. Most recently Guthrie and Napier (1996) have done a comprehensive study that confirms that this quantized effect is present in the redshifts of normal galaxies within the local supercluster ($cz < 2000$ km s^{-1}) and they actually found $c\Delta z = 36$ km s^{-1}. Dr. Napier will give a more detailed discussion of this effect in the following paper. While the effect is small and does not detract from generally accepted views that the bulk of the redshifts of galaxies is due to expansion, it remains totally unexplained.

I now turn to the quasi-stellar objects, where as I shall show that not only is the largest component of the measured redshifts almost certainly of non-cosmological origin, but it also appears to show a remarkable numerical periodicity.

2 Quasi-stellar objects (QSOs)

The QSOs were discovered in 1960 from a combination of radio and optical observations (for details see the monograph of Burbidge and Burbidge 1967). They were originally discovered as compact, non-thermal radio sources that are optically indistinguishable from faint blue stars, but they have very large redshifts. It soon became apparent that the majority of the QSOs are radio quiet.

Since cosmologists have always believed that the redshifts must either be due to the Doppler effect or to cosmological expansion, and since most of them were not concerned about the physics of the objects, they immediately assumed that they could use QSOs as cosmological probes. Within the first year of the discoveries a redshift z of order 2 had been found for the radio QSO 3C 9, and it was immediately obvious from the very few redshifts known then that there was practically no correlation of apparent brightness with redshift. As the number of redshifts increased the absence of anything other than possibly a very weak correlation was very clear. Put more starkly, if the QSOs had been discovered first we would never have concluded that the Universe is expanding.[1] Thus, from the point of view of the redshift-apparent magnitude relation there is no prima facie evidence that the redshifts are due to expansion.

Of course, the absence of a correlation does not mean that the redshifts are not cosmological, but it means that if they are, there is a wide range of luminosity of these objects at every redshift. Next, it was established that the optical and radio flux from the QSOs varies in time, something that was unheard of in galaxies of stars. It means that the radiating regions of the QSOs are very small – probably no larger than the Solar System. Hoyle and I immediately saw that this placed severe limits on the properties of the radiating process, and with Sargent we showed that either highly relativistic motion of the radiating surfaces was required, or that these objects were much closer to us than would be deduced if it were assumed that their redshifts are due to the expansion of the Universe (Hoyle, Burbidge, and Sargent 1966a, b).

---

[1] Of course, it can be argued that a good Hubble relation can only be obtained by using "standard candles," which at large redshifts are luminous galaxies, or supernovae of Type Ia. However, it should be remembered that Hubble (1929) first found the expansion relation using field galaxies, i.e., whatever was available, and Humason, Mayall, and Sandage (1956) got a good result for ∼900 field galaxies.

A counter argument was, and is, based on continuity. Since the nuclei of active (Seyfert) galaxies have similar optical properties to QSOs, it was suggested, first by Kristian (1973) and more recently by many others that QSOs are simply the active nuclei of giant elliptical galaxies at high redshifts. In my view this argument is not convincing because there are very few, if any, high redshift QSOs with optical "fuzz" around them in which it has been conclusively demonstrated by spectroscopy that this "fuzz" is starlight with the stars having the same redshift as the QSO, and there are clearly many high redshift QSOs very close to galaxies like M82 where there is no fuzz.

In 1966 Hoyle and I wrote a detailed paper (Hoyle and Burbidge 1966) discussing whether or not the evidence suggested that the QSOs lie comparatively nearby, or whether they lie at cosmological distances. We concluded that there was evidence on both sides, but we suggested that if the bright QSOs were local, they were probably ejected from comparatively nearby galaxies. Soon after this, it became apparent from statistical arguments (cf. Burbidge *et al.* 1971 and others) and the work of Arp (cf. Arp 1967, 1987), and others (cf. Hoyle *et al.* 2000, Chapters 11 and 12) that a number of the bright QSOs with high redshifts are very closely associated with bright galaxies with very small redshifts. The conclusion was that they are physically associated.

Thus by the 1970s there were many observations that suggested that some QSOs, and thus a fraction of the compact radio sources, do not lie at cosmological distances, and thus cannot be used for cosmological studies.

In the nearly 40 years since then, the observational evidence supporting this view has grown, but unfortunately nearly all of the leading cosmologists and most astrophysicists have ignored this result. Many still refuse to accept the evidence, continuing to doubt the statistical arguments or even the more compelling evidence of luminous connections, though we believe that by now it is overwhelming. For others there has apparently been such a widespread conviction that the big-bang model is correct that it is supposed that such results supporting the idea that at least some QSOs have non-cosmological redshifts can be disregarded, though, in fact, cosmology and the physics of QSOs are two topics that can be uncoupled. The other major problem is, of course, that we have no ready explanation of non-cosmological redshifts. But it is these phenomena, and others discovered in high-energy astrophysics, that have determined the direction of research of some of us since then. The new observations started with the identification of the extragalactic radio sources with very active (explosive) centers of galaxies – now generally called active galactic nuclei (AGN). They and the QSOs have shown without question that there are major sources of energy and ejected mass in the Universe that appear sporadically in a wide variety of galaxies long after these galaxies formed. This far, of course, everyone has been prepared to go. But it is a far cry from the idea that galaxies and their precursors all formed in the very early Universe. The connection that is normally made is that all of the activity is due to the infall (accretion) of matter on to massive black holes in the centers of galaxies.

However, in the 1950s and 1960s, V. A. Ambartsumian had already made the radical proposal that the centers of galaxies are places where the material of *new* galaxies is created and ejected. While Ambartsumian's ideas, based completely on observations, have been largely ignored by the cosmological establishment, which wants to believe that everything arose in an early Universe, these are the cosmogonical ideas out of which, in the 1990s, Hoyle, Narlikar, and I formulated the quasi-steady-state cosmology (QSSC) in which it is argued that the centers of active galaxies *are* the creation sources, and it is in them, in the near vicinity of black holes, that the C (Creation)-field operates. Thus matter is being created out of a set of singular points associated with the nuclei of galaxies. Thus, using biblical terminology, galaxies do beget galaxies. This leads to expansion and contraction with a period of about 40×10^9 years superimposed on an overall expansion with a characteristic time $\simeq 10^{12}$ years. This is a cyclic universe, which does not contract to extremely small dimensions (Narlikar and Burbidge 2004).

Thus the picture that we have is that QSOs that are being ejected from the nuclei of active galaxies eventually evolve into young galaxies – this is Ambartsumian's cosmogony as it appears to be operating.

3 The redshifts of QSOs

The most difficult unsolved problem that we have to deal with is associated with the nature and the distribution of the redshifts of the QSOs. Over the last 40 years many surveys of QSOs have been made, and the total number that have been identified with measured redshifts is currently about 50 000 (Veron and Veron 2003). These together with active galaxies whose emission line spectra are usually indistinguishable from those of QSOs have the redshift distribution shown in Fig. 16.1, which has been obtained from the most recent catalog of Veron and Veron (2003). A total of about 65 000 objects is contained in the histogram.[2] The redshifts range from very small values corresponding to many AGN at low redshifts, to a very small number of QSOs with redshifts $\simeq 6$. In Fig. 16.1 the majority of the objects with $z < 0.5$ are classified as active galaxies (AGN). For very low redshift systems the outer parts of galaxies *can* be seen, but in general for most objects classified as AGN only the nucleus has been detected, and it is *assumed*, but frequently *not proven*, that galaxies of stars are present. The main characteristic of all of these objects is that for the AGN nuclei, and for the QSOs, the spectra in the ultraviolet and

---

[2] I am indebted to Kate Ericson of the San Diego Supercomputing Center for making the plot and compiling the frequency distribution as a function of z.

THE DISTRIBUTION OF REDSHIFTS OF QSOs AND AGN COMBINED
(Plot due to K. Ericson based on the catalogue of Veron & Veron (2003))

Bins	Frequency
0	8957
0.2	4950
0.4	3782
0.6	4017
0.8	4250
1	4787
1.2	5101
1.4	5387
1.6	5218
1.8	5183
2	4560
2.2	3322
2.4	1791
2.6	1107
2.8	615
3	437
3.2	230
3.4	92
3.6	185
3.8	128
4	147
4.2	122
4.4	71
4.6	34
4.8	22
5	7
5.2	1
5.4	2
5.6	1
5.8	2
6	1
6.2	3
6.4	0

Total	64,512
> 1	38,556
> 2	12,880
> 3	1,485
> 4	413
> 5	17
> 6	4

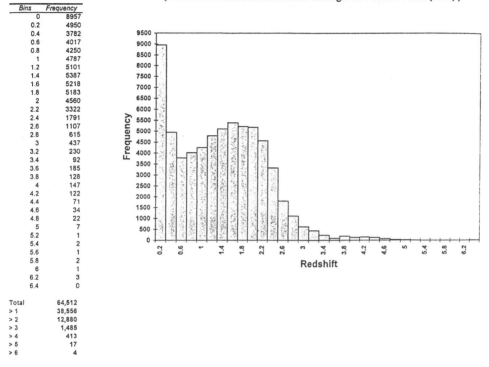

Figure 16.1 This is a histogram containing all of the measured redshifts of AGN and QSOs from the catalog of Veron and Veron (2003).

optical are dominated by broad strong emission lines superposed on a continuum of non-thermal origin.

Beyond about $z \simeq 0.5$ the vast majority of all of the objects in Fig. 16.1 are QSOs. While there are only about 50 000 redshifts in this plot it is assumed that they are representative of the redshifts of QSOs distributed uniformly over the whole sky. Based on the surface density measured in small areas we expect that about 10^6 QSOs down to $m_v \simeq 20$ could be found if the whole 4π steradians was surveyed.

While the numbers of QSOs at very low redshifts are exceedingly small, they increase steadily until a maximum is reached close to $z \sim 2.2$, and then the numbers fall off precipitously so that beyond $z \sim 3$, the QSOs become increasingly very rare. Osmer, Schmidt, and others have shown that this is a real effect and not a result of observational selection. However, as the optical spectrum is shifted more and more into the red it is harder and harder to identify QSOs and those at the highest redshifts have only been detected by using special techniques. In this latest catalog we see from Fig. 16.1 that there are only about 400 QSOs or $\sim 0.8\%$ of the total, with $z > 4.1$. If all of these redshifts are cosmological in origin, as the majority of astronomers still believe, this distribution must be interpreted as due to an evolutionary effect as the Universe has evolved, and there is extensive literature on this topic pioneered by Maarten Schmidt and others.

4 Periodicity in the QSO redshift distribution

If, on the other hand, the QSOs are ejected from galaxies as much observational evidence shows, a large part of the redshifts must be intrinsic. In general, the observed redshift, z_0, is made up of three terms, z_c is the cosmological term (this will be the redshift of the parent galaxy), z_d is the Doppler term (which may be positive or negative and represents the line of sight velocity associated with the ejection) and z_i is the intrinsic term, which is associated with the basic physics of QSOs. Thus

$$(1 + z_0) = (1 + z_c)(1 + z_d)(1 + z_i)$$

In the 1960s, as the number of redshifts of QSOs increased, it became clear that there are peaks in the redshift distribution. The first peaks were found in 1967–68 (Burbidge and Burbidge 1967, Burbidge 1968) at $z_0 = 1.955$ and $z_0 = 0.061$, and by the late 1970s (cf. Burbidge 1978) it was clear that there were redshift peaks at $z_0 = 0.061, 0.30, 0.60, 0.96, 1.41$, and 1.96. Karlsson (1971) showed that the peaks are periodic with $\Delta \log (1 + z_i) = 0.089$. It is now clear that the periodicity only occurs in the intrinsic redshift components (z_i).

From the expression for z_0 given above it is clear that the periodicity associated with z_i will become apparent only when:

(a) z_c and z_d are both very small, so that the term involving z_i dominates, or
(b) z_d is very small, and z_c is known so that z_i can be calculated.

What this means in practical terms is that if the periodicity is seen directly in z_0, the QSOs must be associated with galaxies that have $z_c \ll 0.01$.[3] Alternatively, we must be able to measure z_c for the parent galaxy and make the correction. Thus the fact that the early results (cf. Fig. 16.2) involving bright radio-emitting QSOs shows peaks means that all of these QSOs must have come from galaxies with very small values of z_c.

Using various samples of QSOs, I show in Figs. 16.2 and 16.3 histograms showing the peaks from different sets of QSOs. These have been taken from the work of Napier, Karlsson, and me (Burbidge 1978, Burbidge and Napier 2001, Napier and Burbidge 2003, and Karlsson 1990). It is important to point out that the

---

[3] The fact that no periodicity is seen in the histogram showing all of the QSOs (Fig. 16.1) shows that there is a wide range of values of z_c, i.e., the QSOs have been ejected from galaxies with a wide range of cosmological redshifts. Thus the periodic terms z_i are diluted and the peaks are smoothed out.

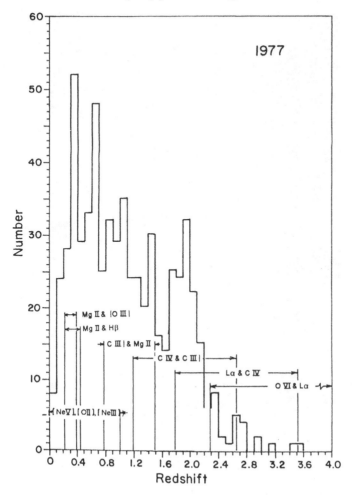

Figure 16.2 This histogram was made in 1977 when only about 600 QSOs had had their redshifts measured (Burbidge 1978). They were predominantly QSOs identified originally as radio sources, they are comparatively bright $< 18.5^m$, and are distributed all over the sky.

peaks beyond $z = 1.96$ at 2.63, 3.44, and 4.45 predicted using the Karlsson formula were found from new samples containing higher redshift QSOs (Burbidge and Napier 2001), which were not identified until long after the samples that Karlsson used to determine the periodicity were found. Needless to say, there have been several attempts to argue from the beginning of this work that selection effects are responsible for what we see, or that the statistical analysis was faulty, or that new samples do not show the effect. All of these issues have been dealt with earlier by Burbidge and Napier (2001) and most recently by Napier and Burbidge (2003). Dr. Napier will discuss them again in the next lecture. It is clear that the periodicity is

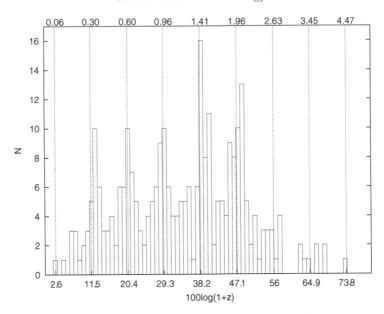

Figure 16.3 This histogram contains QSOs detected since 1990. Included are data from Karlsson (1990) and Burbidge and Napier (2001).

real, and at present we have no explanation for it. It poses what I believe is one of the major unsolved problems in astrophysics.

References

Arp, H. C., 1967, *ApJ*, **148**, 321

Arp, H. C., 1987, *Quasars, Redshifts and Controversies*, (Berkeley, Intersteller Media)

Burbidge, E. M., Burbidge, G., Solomon, P., and Strittmatter, P. A., 1971, *ApJ*, **170**, 223

Burbidge, G. and Burbidge, M., 1967, *Quasi-Stellar Objects*, (San Francisco, W. H. Freeman and Co.)

Burbidge, G., 1968, *ApJ*, **154**, L41

Burbidge, G., 1978, *Physica Scripta*, **17**, 281

Burbidge, G. and Napier, W., 2001, *AJ*, **121**, 21

Guthrie, B. N. and Napier, W., 1996, *A&A*, **310**, 353

Hoyle, F. and Burbidge, G., 1966, *ApJ*, **144**, 534

Hoyle, F., Burbidge, G., and Sargent, W. L. W., 1966a, *Nature*, **209**, 751

Hoyle, F., Burbidge, G., and Sargent, W. L. W., 1966a, *ApJ*, **144**, 534

Hoyle, F., Burbidge, G., and Narlikar, J. V., 2000, *A Different Approach to Cosmology*, (Cambridge University Press)

Hubble, E., 1929, *Proc. Nat. Acad. Sci.*, **15**, 168

Humason, M. L., Mayall, N. V., and Sandage, A. R., 1956, *AJ*, **61**, 97

Karlsson, K. G., 1971, *A&A*, **13**, 333

Karlsson, K. G., 1990, *A&A*, **239**, 50

Kristian, J., 1973, *ApJ*, **179**, 61

Minkowski, R., 1960, *ApJ*, **132**, 908
Napier, W. and Burbidge, G., 2003, *MNRAS*, **342**, 601
Narlikar, J. V. and Burbidge, G., 2004, *Proc. Roy. Soc. A* (submitted)
Veron, M. P. and Veron, P., 2003, *A&A*, **412**, 399

Discussion

Q : J. MORET-BAILLY :

The fundamental constant for the redshifts $\Delta z = 0.062$ has a purely spectroscopic origin. It is deduced from the study of a UV continuous spectrum in atomic hydrogen. The redshift is produced by a transfer of energy from light modes having a high Planck's temperature to colder modes that are blueshifted (these modes are generally thermal, they may be 2.7 K, or close to bright, much redshifted objects, at a higher temperature, and attributed to hot dust). More will be said in the final general discussion.

A : G. B. :

Perhaps your scheme might work. But the problem is that the spectra in which we see the anomalous redshifts are mostly simply interpreted as coming from normal abundances of common elments in a hot gas similar to that which we see in many situations in our own Galaxy and in nearby galactic nuclei containing non-thermal sources.

The atomic physics is quite normal. It is only the redshifts that are so peculiar.

Comment : J. SULENTIC :

In connection with comments on the Lyman alpha forest and the importance to the standard model, caution is needed in using these data. The most recent correlation between the number of absorbers and the source redshift is too good (it appears virtually noiseless!).

Comment : J. SURDEJ :

It is surprising that the redshift peaks reported by G. Burbidge in QSO surveys turn out to correspond to the position of broad emission lines redshifted in standard broad band filters.

For instance, $z =$	1.955	corresponds to	$Ly\alpha$ in the	U band
	0.30	corresponds to	MgII	U band
	0.96	corresponds to	[CIII]	U band
	0.60	corresponds to	MgII	B band
	0.41	corresponds to	[CIII]	B band

Comment : H. ARP :

(In reply to Surdej's comment) There have been numerous tests of redshift period-icity where the quasars are selected by radio or X-ray criteria. There is no optical selection criterion; yet the periodicity is very accurately confirmed. For example the 3C radio sources have been completely identified now and there are about 50 quasars.

Comment : J.-C. PECKER :

Depaquit (together with Vigier and I) have studied (1974, published) the selection effects suggested by Surdej; they do not affect the Karlsson periodicity, as displayed by Burbidge.

Comment : B. CARTER :

How sure can you be that the frequency peaks in the $\ln A$ plots are statistically significant, not just noise?

Q : M. FROISSART :

If one thinks of these periodicities as components of velocities, how do the other components behave, supposing that the Universe is isotropic?

A : G. B. :

I cannot see how the anomalous redshifts can be understood in terms of velocities, i.e., Doppler shifts, because they are all redshifts with no blueshifts.

17

Statistics of redshift periodicities

W. M. NAPIER

Cardiff University, 2 North Road, Cardiff CF10 3DY

Abstract

Claims that ordinary spiral galaxies and some classes of QSO show periodicity
in their redshift distributions have been investigated using high-precision data and
rigorous statistical procedures. The periodicities are broadly confirmed. They are
easily seen by eye in the data sets. Observational, reduction, or statistical artefacts
do not seem capable of accounting for them.

1 Introduction

"Anomalous redshift" claims have appeared in the literature for about 30 years
now and are associated with a few astronomers such as H. Arp, the Burbidges,
and W. G. Tifft. The claims are controversial and the author has been engaged in a
long-term project to examine them objectively. Probably the easiest to test are the
claims of redshift periodicity. The search for periodicity in noisy data has a large
literature and is a well-understood process. Three such claims have so far been
examined, namely the 72 km s^{-1} periodicity (in the Coma cluster), the 36 km s^{-1}
galactocentric periodicity (in wide-profile field spirals), and the periodicity 0.089
in $\log_{10}(1 + z)$ (in the redshifts of QSOs close to bright, nearby spirals).

The approach in all cases has been the same: To use high-quality redshift data,
not previously used in formulating the hypothesis, and rigorous statistical methods.
Modern computing power now allows one to generate large numbers of synthetic
data sets with which the real data can be compared. Here I describe the overall
approach and results rather than the technicalities. The latter can be found in papers
in the reference list.

2 The galactic periodicities

The initial claim made by Tifft was that the redshifts of galaxies in the Coma
cluster show a periodicity 72 km s^{-1}, a result that makes no sense in a system

where the observed radial velocities (presumably virialized) have a dispersion of 1000 km s^{-1}. A second major claim, made by Tifft and Cocke (1984), was that there exists a global, galactocentric quantization of redshifts. For galaxies with narrow HI profiles a periodicity of 24.2 km s^{-1} was claimed, while for broad HI profiles the claimed periodicity was 36.2 km s^{-1}. To see these latter periodicities it was necessary to correct for a solar motion of

$$V_\odot = 233.6 \, \text{km s}^{-1}, \, l_\odot = 98.6°, \, b_\odot = 0.2°$$

close to the galactocentric solar motion.

Subsequently, a series of progressively higher periodic frequencies have been claimed by Tifft, some of which lie on the wrong side of the Nyquist frequency. The current study is concerned only with these initial claims, however. An immediate problem arises from the fact that, in correcting for the vectorial solar motion, three free parameters have been introduced. Thus a Hubble flow of, say, 72 km s^{-1} in a system of galaxies with characteristic projected separation 0.5 megaparsec, has a characteristic velocity separation 36 km s^{-1} and we can imagine that this might be made to appear periodic with some "parameter tweaking," in essence hunting for periodicity. The way to handle this problem is to construct synthetic data sets, identical in all respects to the real one except for the periodicity under test, and to operate on them all, real and synthetic, in identical fashion.

The 72 km s^{-1} claim was tested by Guthrie and Napier (1990) using 48 spiral galaxies in the Virgo cluster, which avoided the core and which had well-determined redshifts (formal accuracies $\sigma \leq 10 \, \text{km s}^{-1}$). This is the nearest rich cluster of galaxies, and had not previously been used to test for redshift periodicity claims, and so is an unbiased sample. Guthrie and Napier (1990) attempted to correct these redshifts for infall towards the Virgo cluster and found that there was indeed a strong periodicity of 71 km s^{-1}, essentially identical to that claimed for the Coma cluster.

In Fig. 17.1 the differential redshifts of the 48 spirals are plotted in the fixed, galactocentric frame of reference. For this plot the latter was taken to be the IAU-approved

$$V_\odot = 220 \, \text{km s}^{-1}, \, l_\odot = 90.0°, \, b_\odot = 0.0°$$

where ($V_\odot$, $l_\odot$, $b_\odot$) are respectively the speed, galactic longitude, and galactic latitude of the Sun's velocity vector around the nucleus of the Galaxy. The data in Fig. 17.1 have been smoothed by a standard procedure: The data set is converted from the velocity to the frequency domain, high frequencies are chopped off, and the remaining signal is reconverted back to the original velocity domain. It is assumed that the high frequencies so removed correspond to noise and measurement error

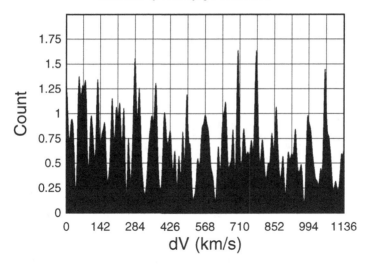

Figure 17.1 Differential redshifts dV of 48 spiral galaxies in the Virgo cluster. Data are plotted in the galactocentric frame of reference as described in the text, and a smoothing with cutoff 13 km s^{-1} has been applied corresponding to the rms sum of the formal redshift uncertainties. The vertical lines are the best-fit periodicity of 71 km s^{-1}. The periodicity is illustrated out to 1000 s^{-1} but extends out to d$V = 3000$ km s^{-1}.

and in the present case the cut-off was taken as 13 km s^{-1} corresponding to the RMS sum of the measurement errors.

A periodicity is obvious to the eye, and routine power spectrum analysis shows it again to be $\sim$71 km s^{-1}. Allowing for several freedoms (excluding dwarf irregulars from the study, arbitrariness in defining the "core," etc.), the periodicity is found to be significant at a confidence level $\sim$10^{-4}. Since the differential solar motion correction across the few degrees subtended by the Virgo cluster is small, uncertainties in this adopted solar apex are second order. In fact the periodicity is observed strongly over a very wide range of solar vectors encompassing the galactocentric one.

Guthrie and Napier (1996) then tested the $\sim$36 km s^{-1} claim for wide-profile spiral galaxies, culling high-precision redshift data from the Bottinelli *et al.* (1990) catalog, simulations having indicated that for the sample sizes employed the effect would only be seen in data of the highest quality. After excluding data employed by Tifft and Cocke (1984), there remained 97 spirals with redshifts measured formally to $\sigma \leq 3$ km s^{-1}. Each redshift had been measured and reduced by at least five groups of observers using five different radio telescopes. These were generally galaxies with broad HI profiles. A remarkably strong periodicity $\sim$37.5 km s^{-1}, again very close to that claimed, does indeed emerge for vectors in the neighborhood of the solar motion (Fig. 17.2). Significance testing again involves the creation of

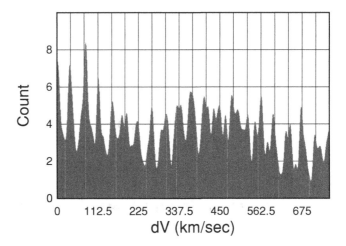

Figure 17.2 Differential redshifts dV of 97 spiral galaxies in the Local Supercluster with systemic redshifts measured to $\sigma \le 3$ s^{-1}. Data are plotted in the frame of reference $V_\odot = 216$ km s^{-1}, $l_\odot = 96°$, $b_\odot = -11°$ and smoothed with a cutoff 13 s^{-1}. The vertical lines are the best-fit periodicity of 37.5 s^{-1}. The first 21 cycles are shown, but in fact the periodicity is detectable out to at least 90 cycles within the LSC.

synthetic data sets and exploration of the power they generate by chance. The hypothesis of non-periodicity is thereby rejected at a significance level $\sim 10^{-5}$.

In this case, since the galaxies are scattered over the sky, individual corrections for the solar motion vary considerably and so the peak powers generated vary strongly with the adopted solar apex. A real periodic signal occurring in the frame of reference of a single velocity generates signals at other velocity vectors and identifying the "real" signal is not a trivial exercise. To find the "true" solar vector, the sensitivity of the signal to variations in the vector must be decreased. This can be done in a number of ways, but the following approach is particularly instructive. About half the galaxies in the sample belonged to small groups or associations containing two to six companions. By looking for periodicity in the *differential* redshifts within these groups, the sensitivity of the signal to $V_\odot$ may be decreased. The resulting data set is small (50 galaxies) and so was enhanced by adding galaxies obtained from a catalog by Tifft (1976) with measured signal-to-noise ratios greater than 10 and which also belonged to cataloged groups. This enhanced data set contained 80 galaxies in 28 groups scattered throughout the Local Supercluster. The power distribution in these 28 LSC groups turns out to have a well-defined maximum at ~ 37.5 km s^{-1} for a solar vector (remarkably!)

$$V_\odot = 220 \text{ km s}^{-1}, l_\odot = 90.0°, b_\odot = 0.0°$$

It is tempting to assume that the $\sim$72 km s^{-1} periodicity found in the Virgo cluster is a harmonic of the $\sim$37.5 km s^{-1} one found for field galaxies. After all, a mere 48 spiral galaxies were employed in its derivation, spread over a range of $\sim$3000 km s^{-1}, leaving most "quanta" unoccupied; one can imagine that the higher frequency oscillation might go undetected, with the power spectrum machinery settling on a harmonic. However, extensive trials by the author involving synthetic Virgo clusters with inbuilt $\sim$37.5 km s^{-1} periodicity have so far failed to yield a false $\sim$72 km s^{-1} one.

Is the $\sim$37.5 kms^{-1} periodicity a local phenomenon, peculiar to individual groups? Or is it global, that is, is there phase coherence from one group to another? This question was explored by constructing synthetic local superclusters. The procedure was to preserve the internal relative redshifts of each group but shift their systemic redshifts bodily by an amount just sufficient to destroy any phase coherence. Trials indicated that the signal in the real LSC is significantly different from those in the synthetic ones. Thus the periodicity is a global rather than a local phenomenon, occurring at least throughout the inner regions of the LSC. A bonus of these simulations is that they make an "artefact" hypothesis hard to sustain: What artefact could produce phase coherence in the galactocentric frame of reference for galaxies widely separated over the sky?

3 A fixed or variable solar apex?

Although the hypothesis of periodicity is preferred over that of non-periodicity at a high confidence level, it is less clear that the "real" periodicity is with respect to a single, fixed vector. Why should a galaxy 10 megaparsec away care about the Sun's motion around the centre of our Galaxy? Radial motion with respect to a local centroid would seem to be another possibility, but to test this one needs good distance information for galaxies. Karachentsev and Makarov (1996) – hereinafter KM – have determined a running apex for 103 galaxies within 500 km s^{-1} of the Sun. The overlap with the Bottinelli *et al.* data set is small – most of the galaxies do not meet the $\sigma < 3$ km s^{-1} criterion adopted by Guthrie and Napier (1996). Distances were in large part determined photometrically with the 6-m telescope.

Galactocentric velocities are plotted against distance in Fig. 17.3. It has slope 61.1 $\pm$5.0 km s^{-1} and intercept -4.7 ± 18.2 km s^{-1}. This intercept is remarkably small: Either the expansion of the Universe is centered on the nucleus of our Galaxy (!), or this is a manifestation of the anomalous quiescence of the local neighborhood, remarked on by Baryshev *et al.* (2001), and others. "Anomalous" is in relation to expectations from CDM N-body simulations, which predict order-of-magnitude

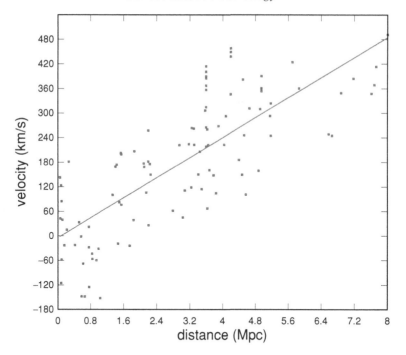

Figure 17.3 Velocity–distance diagram for 103 galaxies within 500 km s^{-1}, in the galactocentric frame of reference.

higher velocity dispersions for Local Groups, with or without biasing (Governato *et al.* 1997).

The remarkably small dispersion ($\sim$72 km s^{-1}, as it happens) of the residuals in relation to the running apex was also remarked on by KM, who tabulated the velocity residuals, but for some reason did not plot them. This deficiency is remedied in Fig. 17.4, which reveals clear evidence for structure (the residuals are plotted both raw and smoothed). A value $\sim$15 km s^{-1} was taken for the high-frequency cut-off, although it is difficult to estimate. Peaks are clearly evident and are consistent with a periodicity $\sim$36 km s^{-1}, relative to the running apex.

Figure 17.5 shows the outcome of a search for periodicity in the KM data using a *fixed* solar vector. It turns out that there is a periodicity 35.2 km s^{-1} when one subtracts out a velocity component corresponding to an apex

$$V_\odot = 220 \, \text{kms}^{-1}, l_\odot = 100.0°, b_\odot = -17.0°$$

This again illustrates one of the problems in analyzing the phenomenon: While the periodicity is readily observable in high-quality data sets, it is not always clear which vector is "real," and which are "ghosts" or "harmonics," or even whether a single, fixed vector is involved.

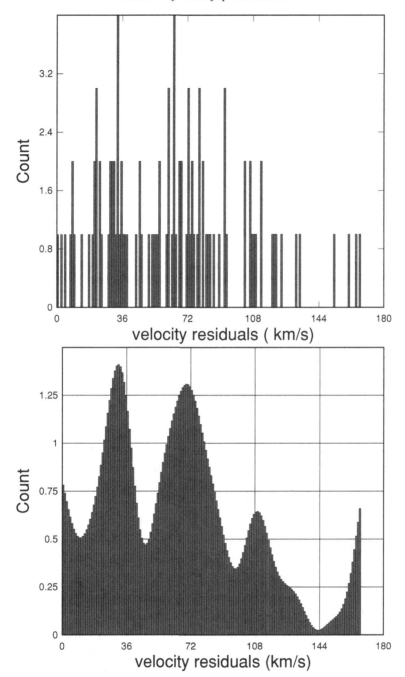

Figure 17.4 Distribution of velocity residuals for 103 KM galaxies relative to a running solar apex out to $500 \, \text{km s}^{-1}$. Upper graph: The raw data. Lower graph: The distribution has been smoothed by removing high-frequency noise corresponding to a cut-off at $15 \, \text{km s}^{-1}$.

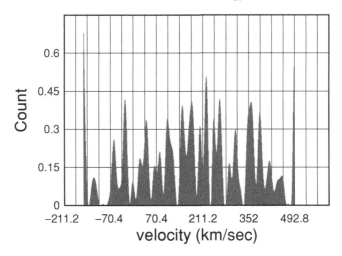

Figure 17.5 Velocities of 103 KM galaxies relative to a fixed solar apex as described in the text. Smoothed with a cut-off at 15 km s^{-1}. There is a best-fit periodicity of 35.2 km s^{-1}.

4 The QSO periodicity claim

The third anomalous redshift claim tested so far is that QSOs in the neighborhood of bright, nearby, active spirals show a periodicity of 0.089 in $\log_{10}(1 + z)$. There is some imprecision in the formulation of this hypothesis (what do we mean by close? By active? What phase should we associate with this periodicity, and with what standard errors?). The hypothesis was tightened up somewhat by bootstrap sampling of 116 QSOs used by Karlsson (1990) to test it, and fresh data were then employed as described in Burbidge and Napier (2001), alias BN: These comprised 57 QSO pairs with separations less than 10 arcseconds, 39 X-ray QSOs near active galaxies (comprising a complete sample), and 78 3C(R) radio QSOs, again comprising a virtually complete sample. Figure 17.6 is a histogram of the combined Karlsson and BN data sets: The periodicity is clearly present and seems to extend three cycles beyond that originally claimed. Monte Carlo trials yield a formal significance level of a few parts in 100 000, whether the null hypothesis is defined through smoothing of the given data or the z-distribution of QSOs as a whole.

It has often been argued that the QSO periodicity is an artefact of observational selection effects (see the discussion in BN). However the data employed here were selected precisely to avoid such effects. It was also claimed that the result is a statistical artefact caused by edge effects (Hawkins *et al.* 2002), but this has been shown to be erroneous (Napier and Burbidge 2003): *inter alia*, edge effects were automatically allowed for in the procedures employed, and the periodicity is easily seen by eye, without any statistical analysis (Fig. 17.6). Again, however, although the periodicity is clear, the circumstances under which it arises are not. For example,

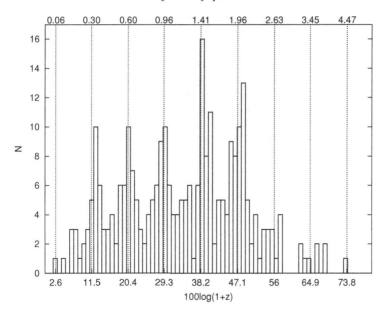

Figure 17.6 Redshift distribution of 290 QSOs compiled by Karlsson (1990) and Burbidge and Napier (2001) in order to test claims of a periodicity 0.089 in $\log_{10}(1 + z)$. Selection procedures, etc. are discussed in BN.

the Karlsson QSOs are for the most part radio loud. Likewise the QSOs in the BN data set tend to be noisy. It remains to be seen whether proximity to bright spirals really is the determining factor.

5 Discussion

The periodicities are empirical findings and are neutral about, say, the cosmological or local provenance of QSOs. It is interesting that a $\log(1 + z)$ periodicity is predicted in vacuum-dominated cosmological models, and an oscillating expansion is also expected in QSSC.

The overall structure of our neighborhood, out to at least 10 megaparsec, is one in which a fractal distribution of galaxies expands with remarkable linearity and coldness, with a redshift periodicity superimposed on the expansion. Current CDM models cannot explain these features. It remains to be seen whether they could be modified to do so, or whether one needs to think outside the box altogether.

References

Baryshev, Yu. V., Chernin, A. D., & Teerikorpi, P., (2001). *Astron. Astrophys.*, **378**, 729.
Bottinelli, L., Gouguenheim, L., Fouqué, P., & Paturel, G., (1990). *Astr. Astrophys. Suppl.*, **82**, 391.
Burbidge, G. & Napier, W. M., (2001). *Astron. J.*, **121**, 21.

Governato, F., Moore, B., Cen, R. *et al.*, (1997). *New Astr.*, **2**, 91.

Guthrie, B. N. G. & Napier, W. M., (1990). *Mon. Not. R. Astr. Soc.*, **243**, 431.

Guthrie, B. N. G. & Napier, W. M., (1996). *Astron. Astrophys.*, **310**, 353.

Hawkins, E., Maddox, S. J., & Merrifield, M. R., (2002). *Mon. Not. R. Astr. Soc.*, in press.

Karachentsev, I. D. & Makarov, D. A., (1996). *Astron. J.*, **111**, 794.

Karlsson, K. G., (1990). *Astron. Astrophys.*, **239**, 50.

Napier, W. M. & Burbidge, G., (2003). *Mon. Not. R. Astr. Soc.*, **342**, 601.

Tifft, W. G., (1976). *Astrophys. J.*, **206**, 38.

Tifft, W. G. & Cocke, W. J., (1984). *Astrophys. J.*, **287**, 492.

18

Local abnormal redshifts

Jean-Claude Pecker

Collège de France, Paris, France

Abstract

Observations of "abnormal" (non-Dopplerian) redshifts in the spectrum of nearby sources (the Sun, binary stars, close-by galaxies in groups), and of "abnormal light deflection in the vicinity of the Sun," are presented. Emphasis is given on the need of reconsidering the observations, which have not been seriously considered since the 1970.

During the early 1970s, Chip Arp started discovering several cases of "abnormal" (i.e., non-Dopplerian) redshifts in the spectra of extragalactic objects. It is one of the most important observational discoveries of our times, in my opinion. At about the same time, I became interested in the abnormal redshifts found in the spectrum of the Sun. J.-P. Vigier, at the same time, was involved in understanding the nature of the photon, along the lines defined by Louis de Broglie, and he did not accept the idea of a zero rest-mass of the photon. We put our efforts together, and we tried to link the abnormal redshifts observed in the local, nearby, universe as consequences of some "tired-light" mechanism, closely linked with the rest-mass of the photon, which we assumed to be a "non-zero restmass," without actually knowing anything else but an upper value of this rest-mass.

I feel it is a need today to remind the audience of these local, solar and others, abnormal redshifts, although they were mentioned extensively several years ago, but they were neither properly confirmed nor really accounted for. Almost all relevant references can be found in our review paper (Pecker 1977).

1 Solar limb redshifts

Several observers (Saint-John 1928, Freundlich *et al.* 1930, Adam 1948, 1959) have noted a redshift of the spectral solar lines at the limb of the Sun, in excess of the prediction of General Relativity.

New measurements were made in the 1960s and 1970s, notably by Roddier 1965 (lines of Sr I), Snider 1972 (KI), and Brault 1962 (NaI); they found, respectively:

$$(\Delta\lambda/\lambda_{meas})/(\Delta\lambda/\lambda_{GR}) = 1.13 \pm 0.05; 1.01 \pm 0.06; 1.05 \pm 0.05$$

Knowing the extreme accuracy of Roddier's resonance spectrograph, we tend to give more credit to his observations than to any other.

So there is an indication for an abnormal additional redshift, to be compared with Einstein's gravitational redshift.

Why?

Our interpretation was that some interaction was responsible for this effect, that of photons emitted by the photosphere with particles encountered around the Sun. Some other theoreticians (Schatzman and Magnan, 1975) assigned the observed effect to some rather improbable (in our views, Jorand, 1962) motions occuring inside the photosphere. But here is not the place to either argue or interpret.

2 Eclipses of a radio source by the Sun

The photons (radio waves) from the source pass, before and after the eclipse, very near the Sun, and crosses the solar coronal layers along a long path. The classical theory is that no redshift should be observed. But the observations, compiled by Mérat *et al.* (1974a), show that a significant redshift affects the lines of the radio source, immediately before and after the total eclipse; this is stronger of course when nearer to the solar limb.

Actually, three different observations, coherent with each other, have been made: The eclipse of the radio source TauA, at 21 cm (H radio-line) (Sadeh *et al.* 1968), that of the maser source W28S (at 18 cm) (Ball *et al.* 1970), and that of the vessel Pioneer 6 (at 13 cm) (Levy *et al.* 1969). The three authors trace the variation of $\Delta\lambda/\lambda$ before and after the eclipse. Depaquit *et al.* (1974) have shown that they were compatible with each other.

It is now suggested that a similar effect is also observed in the recent anomalies of Pioneer 10 and Pioneer 11 signals. They seem to show a deceleration of the vessel. Is it due either to a blueshift (as it could possibly be in the case of a trajectory within the protoplanetary-planetary disk), or to a redshift decreasing with distance in the case where the trajectory would be far enough from the planetary disk?

I again refer to my review of 1977 for the earlier references. The cases of Pioneer 10 and 11 are more recent 2003 observations (NASA Website).

3 Excess of angular deviation of light when near the solar limb

One knows that the effect can indeed be predicted in the framework of Newtonian physics; the calculated value of the deviation, at the solar limb, is 0.88″. The GR reached instead a prediction of 1.75", twice the Newtonian value. Such a value was indeed not really observed in 1919 by the famous eclipse expedition of Eddington; but these eclipse observations were nevertheless good enough to lead the scientific world to adopt the General Relativity. Since 1919, many observations have been piling up, and have confirmed this general conclusion.

However, some observers found a higher value of the deviation near the limb. Mikhaïlov found 2″03 (the analysis of this work was done by Mérat, 1974, its discussion by Mérat *et al.*, 1974b). It was noted long ago that redshift and deviation are closely associated (notably by Fürth 1964, and many other authors since). This excess of deviation over the GR value may be linked with the excess of redshift described above.

4 The case of binaries: A redshift of the "gamma-line"

The radial velocities of binaries display two oscillatory behaviors, each corresponding to the spectra of each of the two components, with the same period and in phase opposition. The gamma-line is defined, for each of these two periodic variations of the radial velocity, as the average radial velocity of this component. Normally, the two gamma-lines coincide. But it happens in some stars that it does not. Then one of the two gamma-lines is affected by some redshift that can be labeled "abnormal," possibly due to the nature of the star concerned. This appears in particular when the star in question is surrounded by a thick atmosphere. Such was the case of the star HD 193576 studied by Wilson in 1940. Struve had also noted this phenomenon in 1944. Kuhi, Pecker, & Vigier (1974) had made a systematic study of a few WR binaries and detected the same phenomenon.

It is difficult to see any classical interpretation of these observations.

5 Nearby galaxies in groups

In several groups of galaxies, containing a bright "mother" galaxy and a few dwarf galaxies, the more compact is the object, the more redshifted is its spectrum (Collin-Souffrin, Pecker, Tovmassian 1974).

Although there are objections (Collin-Zahn, private communication) to an interpretation of this effect by the same type of "mini-bang physics" as the one suggested by Arp or by Hoyle, Burbidge, and Narlikar (Narlikar, this colloquium) for the abnormal redshifts of quasars, we feel that the effect is real, and needs some non-Dopplerian interpretation. Such a type of phenomenon was actually already noted by Bottinelli and Gouguenheim (1973), in the radio domain. We feel that the matter needs further examination and that it should not be forgotten.

Conclusion

All the observed facts briefly mentioned in this paper should be reobserved and rediscussed. But one should remember that there is no one single case of abnormality, but a collection of facts, which seem to indicate, in nearby objects, abnormal, i.e., non-Dopplerian, redshifts. We have given an interpretation, which could neither be confirmed nor refuted by physical laboratory determinations. But efforts should be made to check the reality of these phenomena, and to look for alternative theories to explain in these cases (very different from the "explosive" cases noted in the extragalactic realm) the abnormal observations, if confirmed. If not confirmed, one should explain why these theories were in error.

I would like to quote as a conclusion the final sentence of Einstein's 1916 popular paper on GR.

"One has been furthermore able to derive from this theory a consequence that could be checked by observation, namely a displacement of the spectral lines of the light coming from the largest stars, w. r. to that affecting on Earth the light produced in a similar way, i.e. by the same molecular compound. I have no doubt that this consequence of the GR theory will soon be verified as well as others."

This sentence does refer to the gravitational redshift, but it could be applied as well, as can be shown, to the redshift usually associated with distance, and the properties of the space and matter through which the light has traveled.

References

Adam, M. G., 1948, *M. N. R. A. S.*, **108**, 598
    1959, *M. N. R. A. S.*, **119**, 473
Ball, J. A., Dirkinson, O. F., Lilley, A. E., Penfield, H., Shapiro, I., 1970, *Science*, **167**, 1755
Bottinelli, L., Gouguenheim, L., 1973, *Astron. & Astrophys.*, **26**, 85
Brault, J. W., 1962, Thesis, Princeton Univ.
Collin-Souffrin, S., Pecker, J.-C., Tovmassian, H., 1974, *Astron. & Astrophys.*, **30**, 351
Depaquit, S., Pecker, J.-C., Vigier, J.-P., 1974, *C. R. Ac. Sc. Paris*, **B279**, 559., & **B280**, 113
Einstein, A., 1916, ...
Freundlich, E. F., Brünn, A. von, Brück, H., 1930, *Z. f. Astrophys.*, **1**, 43
Fürth, R., 1964, *Phys. Rev. Letters*, **13**, 221
Jorand, M., 1962, *Ann. Astrophys.*, **25**, 57

Kuhi, L., Pecker, J.-C., Vigier, J.-P., 1974 *Astron. & Astrophys.*, **32**, 111

Levy, G. S., Sato, T., Seidel, B. L., Stelzried, C. T., Ohlson, J. E., Rusch, W. V. T., 1969, *Science*, **166**, 596

Mérat, P., 1974, *Gen. Relat. Gravit.*, **5**, 7567

Mérat, P., Pecker, J.-C., Vigier, J.-P., 1974a, *Astron. & Astrophys.*, **30**, 167

Mérat, P., Pecker, J.-C., Vigier, J.-P., Yourgrau, W., 1974b, *Astron. & Astrophys.*, **32**, 47

Pecker, J.-C., 1977, (review paper), *Colloque UAI n° 37/ CNRS n° 263*, Ed. du CNRS, Paris, p. 451

Roddier, F., 1965, *Ann. Astrophys.*, **28**, 478

Sadeh, D., Knowles, S. H., Yaplee, B. S., 1968, *Science*, **159**, 307

Saint-John, C. E., 1928, *Astrophys. J.*, **197**, 25

Schatzman, E., Magnan, Ch., 1975, *Ann. Astrophys.*, **38**, 373

Snider, J. L., 1972, *Phys. Rev. Letters*, **28**, 853

Struve, O., 1944, *Astrophys. J.*, **100**, 188

Wilson, O., 1940, *Astrophys. J.*, **91**, 379 & 394

Discussion

Q : F. SANCHEZ :
You have described some local non-Doppler observations. Could you speak about the non-Doppler effect observed in the AGN emission oscillations, that we call "cosmic oscillations" because the period 9600,6(1) s enters the *"holographic cosmology"* with a precision of 10^{-4}, and is directly connected with the Balmer wavelength by *c*-free dimensional analysis.

A : J.-C. P. :
Three replies: (1) I have all confidence in Kotov's and associates' observations of the 9600,6(1) periodicities in several spectra (Sun, and others; some of Kotov's papers have been published in the *C. R. Acad. Sc. Paris*). I agree with the fact that some observations having been done from space; it is unlikely that Kotov's periodicity is only due to an effect of resonance with the Earth's rotation period, as it has been claimed. I think Kotov's effect is a real "fact." (2) Can the Kotov's periodicity be considered as "cosmic?" It does not seem to affect all sources; for example it does not appear in the analysis of variable stars light variations. So I still have doubts. (3) As to the relevance in this context of the "holographic cosmology," which you have presented on different occasions, and which have unfortunately not been published, I cannot really comment. I do feel that you found startling coincidences based on a dimensional analysis of various quantities. But an analytical theory, explaining where are these coincidences coming from, seems to me still missing.

Q : J. MORET-BAILLY :
The classical frequency shifts (Doppler, gravitation) taken into account, it remains a blueshift of the radio signals of the Pioneer 10 and 11 probes. The "CREIL"

allows a simple explanation: This effect transfers energy from the beams from the Sun, which have a high temperature (Planck) to the radio signal (and the thermal radiation), which has a low temperature.

The "CREIL" requires a modulation of the beam, which is provided by the noise, which has the same order of magnitude as the signal.

A : J.-C. P. :

You have issued several papers describing the CREIL. In my opinion, it can only explain, if quantitatively developed, rather small redshifts, not those that affect the distant quasars. I would prefer to see the same mechanism explaining the local abnormal redshifts, which I have described, rather than the larger redshifts observed in quasars.

19

Gravitational lensing and anomalous redshifts

J. Surdej, J.-F. Claeskens, and D. Sluse

Institut d'Astrophysique et de Géophysique, Université de Liège,
Belgique
also, Directeur de recherches honoraire du FNRS, Belgique

Introduction

In this chapter, we should like to address the following question: can we invoke gravitational lensing as a possible explanation for anomalous redshifts? In the rest of the chapter, anomalous redshifts refer to redshifts observed for two distinct objects with an angular separation less than $5''$ and whose difference is larger than 0.1.

1 Multiply imaged quasars

Unlike most astrophysical discoveries made during the last century, the physics of gravitational lensing (GL) was understood well before the first example of a multiply imaged object was found (see Einstein 1912 quoted in Renn *et al.* 1997). The existence of multiply imaged, distant sources had been predicted by Zwicky (1937) . . . although the first case of a doubly imaged quasar was only reported in 1979 (Walsh *et al.* 1979). We refer the reader to Surdej and Claeskens (2001) for a recent account on the history of gravitational lensing.

Gravitational lensing coupled with redshift-distance relations has enabled one to make the prediction that cases of multiple images of a distant source with redshift z_s should be detected around a foreground lens with redshift $z_l \ll z_s$.

Following the discovery of the first multiply imaged quasar candidates, some doubt had been cast on the interpretation of gravitational lensing as the possible origin of these systems (see Arp and Crane 1992 for the case of 2237 + 0305). Today (see the URLs http://cfa-www.harvard.edu/castles/ and http://vela.astro.ulg.ac.be/grav_lens) some 92 cases of multiply imaged extragalactic sources have been reported. Among these, the sources and lens redshifts have been successfully measured for 53 of them. All these show so-called anomalous redshifts ($z_l \ll z_s$). Note, however, that not a single case with $z_l \gg z_s$ has been identified. Given this and all the studies that have been successfully carried out for most of these systems (see the non-exhaustive bibliography on Gravitational

Figure 19.1 Setup of the optical gravitational lens experiment.

Lensing in Pospieszalska-Surdej *et al.* 2001; see the URL http://vela.astro.ulg.ac.be/ grav_lens), we are firmly convinced that gravitational lensing coupled with redshift-distance relations may simply account for those apparent anomalous redshifts. In addition to the GL anomalous redshifts, gravitational lensing also predicts that the geometrical configurations observed among the multiple-lensed quasar images are generic. The didactical experiment presented in the next section allows one to visualize most of those expected image configurations.

2 The didactical GL experiment

To simulate the formation of lensed images by a given mass distribution (e.g., a spiral lens galaxy), we are using the optical setup shown in Fig. 19.1.

A compact light source (representing, e.g., a distant quasar) is located on the left-hand side. Then comes the optical lens, which deviates in our case the light rays like an exponential disk (see Refsdal and Surdej 1994 for more details). Behind the lens, we find a white screen with a very small hole at the centre (the pinhole lens). Further behind, there is a large screen on which is (are) projected the lensed image(s) of the source (an Einstein ring plus a very faint central image, in this case) as it would be seen if our eye were located at the position of the pinhole.

In the absence of an intervening galaxy, the large background screen appears to be uniformly illuminated, and the observer only sees the single image of a distant quasar. When setting the spiral lens galaxy perpendicularly to the rays coming from the distant source (see Fig. 19.1), neither the pinhole screen nor the background one are any longer illuminated uniformly. The spiral lens galaxy has redistributed the light in such a way that there is a maximum of light concentration along a bright focal line connecting the source and the lens. There is no single focus, the spiral

Figure 19.2 Intersection of the bright focal line with the pinhole screen. In this case, the mass distribution is axially symmetric and the source, the gravitational lens, and the observer (the pinhole) are perfectly aligned. The resulting image is an Einstein ring, with a very faint central image (see Fig. 19.1).

galaxy acts as an imperfect lens . . . somewhat like a lens affected by spherical aberration. The intersection of this bright focal line with the pinhole screen is seen as a bright spot in Fig. 19.2.

Thus, the maximum amplification is obtained when the pinhole (the observer's eye) is very precisely set on the optical axis, corresponding to the formation of an Einstein ring plus a faint central image. Farther away from the optical axis and in a plane perpendicular to it, the light gets dimmer; the distribution of light does in fact correspond to convergence points there due to three light rays being deflected by the lens, and the total amplification of the images tends towards unity with increasing distance from the axis of symmetry. We may easily observe that if the pinhole lens is set somewhat to the side of the bright spot (see Fig. 19.3a), the Einstein ring breaks into two lensed images, plus a very faint third one, with a typical angular separation equal to the diameter of the Einstein ring (see Fig. 19.3b).

As we may expect, symmetric lenses seldom occur in nature; usually the main lens itself is non-symmetric, or some non-symmetric disturbances are induced by the presence of neighboring masses. In our gravitational lens experiment, the effects of a typical non-symmetric gravitational lens may be simulated by simply tilting the optical lens with respect to the line connecting the source and the lens. The bright focal line along the optical axis that existed in the symmetric configuration then becomes a two-dimensional envelope, called a caustic in optics. A section of this caustic is visible as a closed curve having a diamond shape (made of four folds

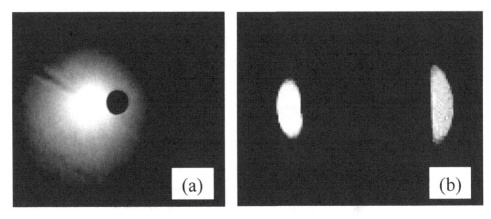

Figure 19.3 As the pinhole is set slightly away from the symmetry axis (a), the Einstein ring breaks up into two images, plus a third very faint central one (b).

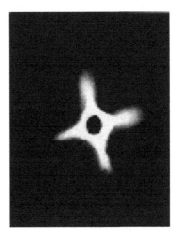

Figure 19.4 Section of the caustic in the pinhole plane for the case of a non-symmetric gravitational lens. The dark spot is the pinhole (see text).

and four cusps) in the pinhole plane, surrounded by an outer elliptical caustic (the latter is not conspicuous in Fig. 19.4).

The word "caustic" in gravitational lensing always refers to this section of the optical two-dimensional caustic (in the symmetric case, the caustic degenerates into a single spot surrounded by a faint outer elliptical caustic, Fig. 19.2). As a result, the Einstein ring that was observed in the symmetric case is now split into four lensed images plus a very faint central one. Depending on where exactly the pinhole lens is located with respect to the caustic, different generic configurations of lensed images are produced. These are illustrated in Fig. 19.5 and compared

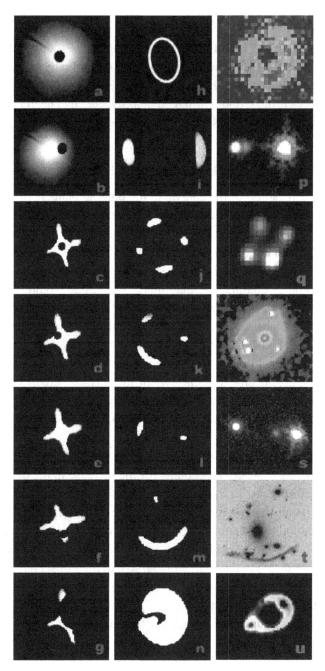

Figure 19.5 Images in the left-hand column represent the light from a distant source that is redistributed over the pinhole screen (in the experiment shown in Fig. 19.1) by a symmetric (a, b) or a non-symmetric (c–g) optical lens and for various positions of the pinhole (observer) seen as a dark spot. The central column (h–n) illustrates the corresponding lensed images projected onto the large screen located behind the pinhole screen, and the right-hand column (o–u) displays known examples of corresponding multiply imaged sources (0047-28078, 1009-0252, H1413+117, PG1115+080, HE1104-1805, Abell 370, and MG1131+0456). Images (p), (r), (s), and (t) were obtained with the Hubble Space Telescope, and the others using ground-based facilities (ESO and VLA/NRAO). From Surdej and Claeskens (2001, ©2001 Kluwer Academic Publishers).

with typical observations of multiply imaged extragalactic sources. All image configurations observed for gravitational lens systems can actually be reproduced with this straightforward gravitational lens experiment. The existence of giant luminous arcs in rich galaxy clusters, doubly imaged quasars, etc., had been predicted before they were actually discovered (Nottale 1988, Refsdal 1964, Liebes 1964).

3 The amplification bias

The brightness distribution of the caustic in the pinhole plane is of course directly related to the amplification of the flux of the background quasar that would be measured by a distant observer. Typically, if the angular distance between the lens and the true source position is less than the Einstein angular radius θ_E, the amplification is expected to be larger than several tens of percent. Considering a flux limited sample of quasars (i.e. magnitude $<m^*$), one would therefore expect to include more cases of multiply imaged quasars than in a volume-limited sample (assuming that the number counts of the quasars at the observed magnitude are large enough; see below). If the average amplification of the flux of background objects (cf. quasars) due to lensing by a population of foreground objects (cf. galaxies or quasars) is A, the enhancement $q(<m^*, A)$ of the surface density of the observed background objects $n_L(<m^*)$ compared to the surface density $n_U(<m^*)$ of unlensed objects, i.e., the amplification bias, is found to be (Narayan 1989):

$$q(<m^*, A) = n_L(<m^*)/(n_U(<m^*)) = n_U(<m^* + 2.5\log(A))/(n_U(<m^*)A) \quad (1)$$

The factor $n_U(<m^* + 2.5\log(A))$ accounts for the increase in the observed number of background objects due to the fact that objects as faint as $m^* + 2.5\log(A)$ can now become brighter than the limiting magnitude m^*, as a result of gravitational lensing amplification by a factor A. However, the objects we observe projected in the lens (sky) plane within a solid angle $d\omega$ are actually located within a smaller solid angle $d\omega_s$ $(= d\omega/A)$ in the source plane, so that their observed surface density must be decreased by a factor $d\omega/d\omega_s$, accounting for the factor $1/A$ in the above equation. If s represents the slope of the logarithmic intrinsic number counts of objects as a function of apparent magnitude m^*, such that

$$s = d(\log(n_U(<m^*)))/dm^* \quad (2)$$

Equation (1) may then be rewritten as

$$q(<m^*, A) = A^{2.5s-1} \quad (3)$$

Depending on the value of the slope s (≥ 0.4 or <0.4), we see that the amplification bias or enhancement factor $q(<m^*, A)$ will be either larger or smaller than 1. For relatively bright quasar samples, it is found that the slope $s > 0.4$ and the resulting amplification bias is larger than 1 (Turner *et al.* 1984, Surdej *et al.* 1993).

3.1 Galaxy–quasar associations

The amplification bias should create an artificial correlation at very small angular separations (a few arcsec) between the high redshift background QSOs and foreground visible galaxies in flux limited samples, even if the former are not multiply imaged (weak lensing). However, the number density of the sources is also diluted by gravitational lensing and Narayan (1989), Kayser and Tribble (1991), Claeskens and Surdej (1998), and many others have shown that the resulting expected overdensity of galaxies in the angular vicinity of QSOs is very low, and cannot reproduce the highest reported values. Claeskens and Surdej (1998) also claim that the comparison is hampered by the small number of statistics and possible selection biases and that about 1500 Highly Luminous Quasars ought to be observed down to a limiting magnitude of $R_{lim} \sim 23$ before a definite conclusion can be drawn.

3.2 Quasar–quasar associations

Burbidge *et al.* (1997) have argued that the observed number of quasar pairs with small angular separations and anomalous redshifts (typically $\Delta\theta < 5''$ and $\Delta z > 0.1$) is not compatible with a random distribution of quasars over the sky. Considering the three such known pairs of quasars (Q1548 + 114 A & B, Q1009–0252 AB & C, and Q1148 + 0055 A & B) in the Véron–Cetty and Véron (2000) catalog of quasars, Sluse *et al.* (2003a) have shown by means of very simple calculations that the probability of finding the three accepted pairs accidentally is of the order of 10%. They conclude that, under realistic hypotheses, the observed number of quasars with anomalous redshifts is not unlikely. They also present arguments showing that gravitational lensing biases are probably not strong enough to significantly increase the expected number of close pairs of quasars with anomalous redshifts. Indeed, in order to get a significant enhancement (typically >2) of the factor $q(<m^*, A)$, a very large average amplification A is needed. For such large amplification factors, strong lensing takes place with the resulting formation of multiple images of the background quasar near the foreground one. The failure to detect with HST a secondary lensed image of the background quasar near the foreground one in these three pairs supports this view (see Fig. 19.6).

4 Gravitational lensing as a predictive tool for lens and source redshift estimates

Considering the several tens of known multiply imaged quasars with known (lens and) source redshifts, observed image configurations, etc., gravitational lensing theories allow one to predict what should be the most likely redshift distribution of the lenses. A reasonably good agreement exists between the observed and the

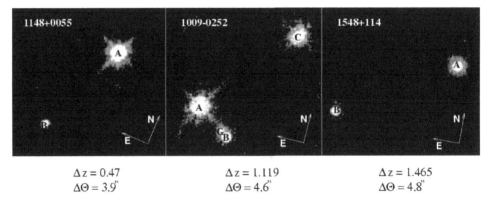

$$\Delta z = 0.47 \qquad\qquad \Delta z = 1.119 \qquad\qquad \Delta z = 1.465$$
$$\Delta\Theta = 3.9'' \qquad\qquad \Delta\Theta = 4.6'' \qquad\qquad \Delta\Theta = 4.8''$$

Figure 19.6 HST F814W observations of the three close quasar pairs with different redshifts known today (Sluse *et al.* 2003, © ESO 2003).

predicted redshift distributions of the main gravitational lenses (Ofek, Rix, and Maoz 2003).

Light rays from a multiply imaged quasar usually sample different path lengths across the deflector. Extinction in the galaxy may thus lead to a differential obscuration and reddening between the observed macro-lensed QSO images. These effects naturally depend on the precise shape of the extinction law and on the redshift of the lens. Jean and Surdej (1998) have shown how accurate photometric observations of multiply imaged quasars obtained in several spectral bands may lead to the estimate of the lens redshift, irrespective of the visibility of the deflector. Jean and Surdej (1998) have estimated that the redshift of the galaxy lens in the system MG 0414 + 0534 was 1.15 ± 0.2, the accuracy depending on the number of broad-band filters and signal-to-noise ratio of the photometry. The spectroscopic redshift for this lens has been subsequently measured to be 0.96 (Tonry and Kochanek 1999).

We should still like to mention that rich galaxy clusters have been used as natural cosmic telescopes to search for very distant objects located behind them. Such searches have been very successful (Franx *et al.* 1997, Mehlert *et al.* 2001).

On the basis of redshifts measured for selected lensed sources seen projected on a rich foreground galaxy cluster, it is possible to constrain its mass distribution and to predict via the cluster gravitational lens model the redshift of extremely faint multiple images for which no redshift has yet been obtained (cf. Ebbels *et al.* 1998). For the case of Abell 1835, Pello *et al.* (2004) have predicted a large redshift for such a source. Subsequent spectroscopic measurements obtained with ISAAC at the VLT seem to confirm their prediction ($z = 10$) for this high-z candidate and attest the validity of the gravitational lensing model coupled with redshift-distance relations to interpret distant views of the Universe.

Finally, for the case of the extended mirage and multiply imaged quasar RXS J1131–1231 (Sluse *et al.* 2003b), it has been possible to reconstruct the source

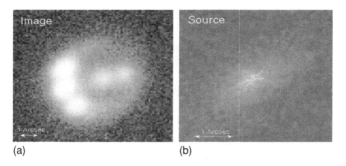

Figure 19.7 (a) The quadruply imaged quasar RXS J1131-1231 ($z = 0.658$) with an optical Einstein ring. (b) "Inversion" of the observed mirage in the source plane. The redshift of the lens is $z = 0.295$.

image, using the lens equation. Adopting the observed redshifts for the lens and the source, and standard redshift-distance relations, the host galaxy of a Seyfert 1 nucleus has been retrieved (see Fig. 19.7). It looks like a spiral galaxy. In our opinion, such image reconstructions via gravitational lensing and standard redshift-distance relations also strongly support the coherence of the adopted model.

Acknowledgements

We would like to dedicate this contribution to the memory of Christian Vanderriest, who was an astronomer at Meudon Observatory, a faithful collaborator, and a very good friend.

We thank very warmly Prof. J.-C. Pecker and J. V. Narlikar for inviting us to participate to this conference. Our research was supported in part by PRODEX (Gravitational lens studies with HST), by contract IUAP P5/36 "Pôle d'Attraction Interuniversitaire" (OSTC, Belgium), and by the "Fonds National de la Recherche Scientifique."

References

Arp, H., Crane, P., 1992, *Phys. Lett. A*, **168**, 6
Burbidge, G., Hoyle, F., Schneider, P., 1997, *Astron. Astrophys.*, **320**, 8
Claeskens, J.-F., Surdej, J., 1998, *Astron. Astrophys.*, **335**, 69
Ebbels, T. M. D., Ellis, R. S., Kneib, J.-P., Le Borgne, J.-F., Pello *et al.*, 1998, *Mon. Not. R. Astron. Soc.*, **295** (1), 75–92
Franx, M., Illingworth, G. D., Kelson, D., van Dokkum, P. G., Tran, K.-V., 1997, *Astrophys. J.*, **486**, L75
Jean, C., Surdej, J., 1998, *Astron. Astrophys.*, **339**, 729
Kayser, N., Tribble, P., 1991, *ASP Conference Series*, **21**, 304
Liebes, S., 1964, *Phys. Rev. B*, **133**, 835
Mehlert, D., Seitz, S., Saglia, R. P., Appenzeller, I., Bender, R. *et al.*, 2001, *Astron. Astrophys.*, **379**, 96–106

Narayan, R., 1989, *Astrophys. J.*, **339**, L53

Nottale, L., 1988, *Ann. Phys. Fr.*, **13**, 223

Ofek, E. O., Rix, H.-W., Maoz, D., 2003, *Mon. Not. R. Astron. Soc.*, **343**, 639

Pello, R., Schaerer, D., Richard, J., Le Borgne, J.-F., Kneib, J.-P., 2004, *Astron. Astrophys.*, **416**, L35-L40

Pospieszalslea-Surdej, A., Surdej, J., Detal, A., Jean, C., 2001, *ASP Conference Series*, **237**, 55–56

Refsdal, S., 1964, *Mon. Not. R. Astron. Soc.*, **128**, 295

Refsdal, S., Surdej, J., 1994, *Rep. Prog. Phys.*, **56**, 117

Renn, J., Sauer, T., Stachel, J., 1997, *Science*, **275**, 184–186

Sluse, D., Surdej, J., Claeskens, J.-F., De Rop, I., Lee, D.-W., Iovino, A., Hawkins, M., 2003a, *Astron. Astrophys.*, **397**, 539–544

Sluse, D., Surdej, J., Claeskens, J.-F., Hustemékers, D., Jean, C., Courbin, F., Nakos, T., Billeres, M., Khmil, S. V., 2003b, *Astron. Astrophys.* 406, L 43

Surdej, J., Claeskens, J.-F., Crampton, D., Filippenko, A. V., Hutsemékers, D. *et al.*, 1993, *Astron. J.*, **105**, 2064

Surdej, J., Claeskens, J.-F., 2001, in *The Century of Space Sciences*, Eds. Bleeker, J., Gliss, J., Huber, M., Kluwer Academic Publishers, pp. 441–469

Tonry, J. L., Kochanek, C. S., 1999, *Astron. J.*, **117**, 2034

Turner, E. L., Ostriker, J. P., Gott, J. R., 1984, *Astrophys. J.*, **284**, 1

Véron-Cetty, M., Véron, P., 2000, *ESO Sci. Rep.*, **19**, 1

Walsh, D., Carswell, R. F., Weymann, R. J., 1979, *Nature*, **279**, 381

Zwicky, F., 1937, *Phys. Rev.*, **51**, 290

Discussion

Q : A. BLANCHARD :

Isn't it true that the statistics of Lyα clouds behind the lens in multiple QSOs should be different from genuine pairs and help to clarify whether QSOs are at cosmological distance?

A : J. S. :

Yes, you are absolutely right! However, cases of multiple images of a lensed quasar with angular separations in the range of arcminute(s) would be necessary in order to carry out this test in a significant way.

Q : J. NARLIKAR :

Do you have cases of absorption lines in the spectrum of a lensed QSO at redshifts corresponding to the lensing galaxy?

A : J. S. :

Yes, many cases are known.

Q : D. ROSCOE :

Is there any possibility that Arp's quasar in NGC 7319 is a lensed object?

A : J. S. :

Significant amplification of a background quasar would necessarily lead to the formation of multiple QSO images, which are not observed. Note, however, that Laurent Nottale and his collaborators have proposed in the past that gravitational lensing amplification could be partly responsible for the observed grouping of galaxies in Stephan's Quintet.

Comment :

H. ARP :

The prime exhibit in the gravitational lens hypothesis is the so-called "Einstein's cross." But instead of having the supposedly lensed background quasars drawn out into small tangential arcs, they are connected back to the central low redshift galaxy by a luminous extension. Moreover, one of these connections was shown by Howard Yee to consist of low density Ly α emission. They are actually physically connected back to this galaxy, which is morphologically too small a mass for the theory in any case. The material in the four quasars is rather ejected orthogonally from the central galaxy, a pattern noted in other cases of quasar ejection. The observations are discussed in my book *Seeing Red*, Arp, Apeiron publ., Montreal, 1998, p 173–175, and references therein.

A : J. S. :

Yes, I have read your paper on the Einstein Cross, but I cannot agree with the conclusions.

Part VII

Panel discussion

20

Panel discussion

Geoffrey Burbidge

University of California San Diego, La Jolla, CA, USA

We had a good discussion of various issues relating to cosmology and there has been a clear division of perceptions of what is considered important evidence. On the one side, the conventional one, we have heard the very detailed evidence of CMBR and high redshift supernovae, evidence that is popularized in the phrase "concordance cosmology." The Universe according to this view went through an inflationary phase, had an era of nucleosynthesis and then had the surface of last scattering when the radiation background became decoupled from matter. The package comes with a large part of the matter energy (around 75%) being dark and hitherto unknown, a substantial part of strange kind of matter (21%) and only around 4% of ordinary matter that we are familiar with. Once you believe all of these ideas, you feel convinced that the cosmological problem is all but solved.

On the other side, some of us have been increasingly worried at what appears to be anomalous evidence, evidence that does not fit into the standard picture just mentioned. Even the very basic Hubble law applied to QSO redshifts seems to be threatened if one takes the evidence on anomalous redshifts seriously. In the 1970s when Chip Arp first started finding such examples, he was told that these were exceptions and that he should find more. He has been doing just that and his cases now include not just optical sources but also radio and X-ray sources. Then there is the evidence of periodicities of redshifts that has not gone away with larger samples. As I discussed, even the gamma ray burst sources appear to show the effect. While there are many things that we do not understand we believe that this cosmogonical evidence fits well into the cyclic universe scheme.

The contrast between the two perceptions gets further highlighted when one notices the large number of speculative concepts that have gone into the standard paradigm: The nonbaryonic dark matter, dark energy, phase transitions at energy well beyond the range tested in the laboratory, etc. These relate to parts of the Universe that will remain forever unobservable and whose physics will remain forever untested in the laboratory. However, without making these assumptions the theory fails. The fact is that we do not know how galaxies form, and for them to form in a big-bang Universe it is necessary to invoke initial density fluctuations and a large amount of nonbaryonic matter to make them condense.

On the other hand, the anomalous evidence ignored by the conventional cosmologists is real, right on our doorstep, and well observable. Surely we need to probe it further and in a way that will enable us to understand if any new physics is needed here. It is unfortunate that the majority of the cosmology community chooses to ignore all of this evidence in the hope that it will go away.

Panel discussion

Blanchard, A.

LATT, UPS, CNRS, UMR 5572, 14 Av E.Belin 31 00 Toulouse, France

During this conference we heard several speakers with very different points of view, in strong disagreement with the so-called standard model[1] and I found this very refreshing, because although I disagree with a large proportion of them, I found it very useful to listen to these different points of view. I also realize that we all follow the same logic, which is the basis of the scientific logic: We first try to understand observations on the basis of standard physics and when we are convinced that some fact does not fit in this scheme we advocate new models, eventually new physics. Of course, different physicists have different ideas on what does fit or not and what should be involved, although I realize during this conference that theoreticians might not have so many new ideas! Indeed, the steady-state theory was based on the assumption of a scalar field leading to an exponential phase of the expansion of the Universe, inflation was based on a scalar field leading to an exponential phase of the early expansion, and quintessence is also a scalar field leading to an exponential phase! My personal vision would be that such an ingredient, which is advocated so often in such different contexts, is probably not the right answer, but this is a personal comment of no scientific value . . . Coming back to the subject of the conference, I realize again that the construction of the big-bang model, which is essentially the primeval atom of Georges Lemaître, is a wonderful construction: Even if one believes it is wrong, it should be acknowledged that it is remarkable that it has been possible to build a coherent picture of what the Universe looks like from the first 10^{-10} s to the present epoch. Saying that, I should add that I have been slightly disappointed by what I hear from Jayant Narlikar, although I am impressed by the intellectual effort to build an alternative cosmology on a very different basis (and having discussed quite often with him also I should say I am impressed by the

---

[1] I should specify what is to me the standard model: It corresponds to a model describing the Universe based on physics as known nowadays. In this respect inflation is not part of the standard model, but rather one possible extension, as it is based on physics at an energy level that is not yet tested. Inflation is a scientific theory in the sense that it can be falsified, and should inflation be falsified the standard model would stay at the same level of scientific success.

fact I always find him to be calm and considered with his contradictors), I found the model that he presented not very exciting because it appears to me that its success was limited to the reproduction – imitation I would say – of the major facts that are naturally explained in the standard big-bang picture, but it seems not natural to me in Jayant's model and the ingredients have to be put in just for that purpose. Of course this is due to the fact that the Universe according to Jayant is not simple, and therefore any fact is not simple to explain. However, the model does not produce any obvious specific prediction that follows directly from the hypothesis on which it is based. That may be a definitive reason why most cosmologists have chosen the standard picture: Because it is based on known physics in conditions that are simple to handle and have led to predictions that have been verified! Indeed, to me a theory is attractive if I can *naturally and in a simple way* use it to explain a large number of different facts, even if this aspect might look somewhat subjective! Certainly the most important aspect is the fact that a theory should make predictions that are verified a posteriori. In this respect the big-bang theory has succeeded remarkably. There is, of course, the example of the spectrum of the microwave sky: It has been clear since its discovery that the model predicts that the spectrum should be a black-body shape. COBE has measured this spectrum to a high accuracy and it is a remarkable spectrum, it is even somewhat surprising that not even a tiny deviation has been found: Things are as simple as they could be! Producing the background radiation with this accurate spectrum by other processes, like dust emission, has not been achieved, at least in the published literature I know. If it is done, I will not be particularly impressed, because I know that the alternative model has been built in detail to reproduce this observation (compare with a model that *predicted* from elementary physical consideration: Thermodynamical equilibrium). François Bouchet has also remembered the properties of the fluctuations of the CMB, the famous C_l, which again were *predicted* before being observed. Similarly, there is a prediction of the polarization properties, which is very specific. Although the big-bang picture could easily accommodate non-standard fluctuations properties, for instance, it might have been possible that the fluctuations have been seeded by topological defects, it remains a very good point for this model that the fluctuations are exactly as they are expected in the simplest picture. There are other non-trivial facts that are obvious in the big-bang picture and that will be terribly "unnatural" in alternative pictures. For instance, G. Burbidge is still defending the idea that QSOs are not at their cosmological distance because they appear to be associated with galaxies. Here we see that different scientists put different weight to observational facts: There are certainly a few cases that are puzzling. If I remember correctly, Jim Peebles carried out a statistical investigation and actually found a surprisingly non-vanishing correlation between galaxies and QSOs, which to my knowledge has not been explained yet (lensing has been advocated, but seems not to lead to an effect strong enough) and yet Jim Peebles firmly advocated in favor of the standard picture! Now what I want to point out is that just after the discovery of QSOs, Refsdal did predict the lensing phenomenon. One should find pairs of QSOs in the sky that are multiple images of the same objects and therefore should have identical spectra. A galaxy should be found in between with a mass that is what is needed to explain the angular separation. Furthermore, Refsdal made the point that if the flux varies with time the second image should show similar variations with a time difference of a few years. All these features have been observed, and are now found routinely! Alternative theories are left with so much to explain . . . There are a couple of other facts (the duration of the SNIa light curve, the change of temperature of the CMB with redshift) that are just simple consequences of the model and that seem to be unnatural to explain in any alternative way. Finally, when I try to find what the weaknesses of the standard big bang are, I get rather more convinced of its robustness! Turning back to alternative cosmological models it is generally assumed that some unknown physics is at the origin of what they consider as an observational fact that the standard picture could not explain. However, for the non-expert I think the choice is still simple: We have on one side a theory (the big-bang picture), which pretends to explain *all* facts relevant to cosmology; on the other side, alternative theories try to reproduce a *couple* – sometimes only one – of these facts and this advocates new physics, the implication often being that if true this would imply the whole standard picture is wrong. Both points of view are logically correct: On one side one acknowledges the agreement with a large set of data, accepting that some aspects are not fitted perfectly well (like for years the age problem – which has now disappeared), on the other side one considers that some observational facts are in such severe conflict with the model that it has to be abandoned.

Still in such a situation, I would say that it is reasonable for the non-expert to trust the standard picture rather than rely on unknown physics that has not yet been agreed by almost all physicists! For instance, I think it was reasonable for the non-expert not to trust immediately general relativity when it was first proposed in 1916! This was entirely new, and exotic. It is only when almost all physicists agreed on this new physics that the non-expert could confidently consider that it has been scientifically established, knowing that science does not describe what is *true* but what *works in reproducing* the world.

Having said that, I cannot resist making some comparisons with the present status of the cosmological constant! According to the above criteria, I realize that I am very close to the logic followed by the defenders of alternative cosmological models . . . Indeed I am one of the very few scientists considering that the existence of the cosmological constant has not been established, beyond reasonable doubts, although the vast majority of the experts in the field would consider the opposite.

On the other hand, the fact that the concordance model requires the introduction of something essentially new in physics, basically a negative gravitational action on long distance, requires that a strong consensus should exist among involved scientists for establishing this conclusion (extraordinary claims need extraordinary evidence). One may consider that this point has not yet been reached, as a large fraction of cosmologists agree that the concordance does reproduce well several major facts, but still that direct evidence for a cosmological constant are too weak. Furthermore, I posit that the observed properties of X-ray clusters as they are known now could not be fitted in any way in the concordance picture (this is at odds with the several current claims on the subject, but I believe our analysis – see my contribution in this book – overcomes all existing previous works). So I posit that unless some of the observed properties of X-ray clusters are strongly biased, the abundance of massive clusters is actually evolving at a level that can be understood only in an Einstein–de Sitter model and therefore that the cosmological constant is small enough not to have any significant contribution to the expansion rate of the Universe. This fact is not yet appreciated at the right level. However, if in the future there are new analyses on clusters' properties, which show that present day properties, as we have obtained them from XMM, are incorrect and that a self-consistent analysis as we did allows the concordance to fit, I will agree that the existence of a non-zero cosmological constant has been established scientifically.

I would like to thank J. Narlikar and J.-C. Pecker for having organized this stimulating conference.

Panel discussion

Michael J. Disney

University of Wales, Cardiff, UK

Rightly or wrongly the majority of our professional colleagues believe in the so-called Concordance Model (CM) of Cosmology. If we are going to undermine their faith in it then I believe we have to follow Karl Popper's dictum and attack its central citadel. One can't hope to win the argument by sapping the outlying fortifications – which could fall and still leave the main fortress intact. No theory in its youth is expected to explain everything satisfactorily. Darwin almost gave up the Theory of Evolution when he found initially that it couldn't explain the parrot's plumage. . . Just because the CM cannot satisfactorily explain some bizarre property of QSOs, is not going to change our colleagues' minds – and rightly so. They are impressed by what it apparently can explain, not depressed by what it presently cannot.

The probability of any hypothesis, that is to say, in Laplacian terms, the degree of rational assent we can attach to it, must rest on the balance between the number of relevant and truly independent measurements it can explain, and the number of parameters it is free to adjust. The number of those independent measurements must comfortably exceed the number of such free parameters because, consciously or not, the hypothesis was initially selected from among an uncountable host of alternatives – precisely because it fitted the early observations. My cosmological friends tell me that the CM contains eight free parameters. Add at least three or four more for the initial model selection, and we have 11 or 12 cosmological measurements we have first to discount. The statistical significance of the hypothesis can then be tested only by the remainder, after this first dozen or so have been subtracted out. Now it is not easy to count the relevant measurements, particularly with regard to their independence, but at my latest attempt I reached only 13 [1]. In other words the significance of the CM rests on at most one or two measurements. The right verdict on the so-called Concordance Model of Cosmology today must surely be: "It fits: but so what?"

This worrying degree of plasticity is reflected in recent history. Supernova observations led to a radical change in one free parameter – the so-called "Cosmological

243

Constant," the BOOMERanG observations to change in another – the curvature – or in this case lack of curvature – of 3-space. It is an unhealthy sign when any hypothesis has to adjust itself radically to fit each new incoming observation. Very unhealthy. It shows that the capacity for predictability of the CM, that is to say its statistical significance, is currently very low.

Turn next to the status of General Relativity (GR), the central theory of modern cosmology. How healthy is that? It was and is consciously and explicitly modeled on Poisson's equation (as one can easily confirm by reading Einstein's original papers) – in other words on the Inverse Square Law of Gravitation – which seems not to work at all at long range. Why else have we invoked so much Dark Matter? The fact that GR seems to work well in the strong field limit is neither here nor there when it comes to cosmology – where it is the asymptotic long-range behavior that is crucial. And as to that we know very little. The mere fact that the Cosmological Constant could be arbitrarily tweaked to fit the Supernova Observations, and yet conflict with nothing else in cosmology, is elegant testimony as to the plasticity of GR at long range. And anyway, much of the aesthetic appeal of GR – i.e., its ability to incorporate curved geometries – has recently vanished with the BOOMERanG evidence that space seems to be accurately flat. One can argue, justifiably I feel, that the theoretical cement holding the foundations of cosmology together is now suspect in itself.

As a panel member for this meeting I've been asked to express some opinions about the "anti-cosmological evidence" – if I may call it so – that we have heard at this conference. So, for what they are worth:

1. I am convinced but confused by the bizarre redshift distributions found by Tifft amongst galaxies, and by Burbidge and Napier amongst QSOs. And so apparently are many of my astronomical colleagues who mutter about them, albeit surreptitiously over their beer. The issue here is surely statistical significance – which I have to admit sometimes looks high. What would be more convincing would be to take those same numbers, put them into a quite different and less emotive context, and ask our statistician colleagues whether they are still significant. After all, significance and context should here be entirely independent. If they still remain significant then all astronomers will have to sit up.

2. As to QSO alignments I am not impressed. To my mind the claims for them rest on selected objects and a-posteriori statistics. Besides, QSOs *do* dwell in galaxies – with HST we can nowadays clearly see many of them in their galaxy homes. Anyway it's time to move on from QSOs. Once upon a time they were the only inhabitants of high-redshift space – and very difficult ones to work with at that. Now we know of thousands of relatively well-behaved galaxies that lie out there, even beyond the QSOs in some cases. What should be worrying cosmologists more is the failure of those same galaxies to fit the Tolman test – i.e., apparent surface brightness should

fall off as the inverse fourth or higher power of (redshift plus 1) – without a very specific adjustment for galaxy evolution. It is the same story over again: The CM can be made to fit – but only if you fudge some auxiliary hypothesis.

In summary, I believe the grounds for refusing to believe in the current cosmological paradigm are very strong. The supporting numerical evidence is underwhelming, while the underlying theory is suspect – and certainly unsupported by the observational evidence where it matters most. And as, from an epistemological point of view, cosmology will always be a suspect subject, the burden of proof lies, and must surely always lie, entirely with its advocates. Just because they themselves are impressed doesn't mean we have to be too. As the historian Daniel Boorstin said [2], "The great obstacle to discovering the shape of the earth, the continents, and the ocean was not ignorance but the illusion of knowledge." Or as Josh Billings once put it in a more homely way: "Tain't what a man don't know as makes him a fool. It's what he do know as just ain't so."

I am increasingly convinced that in this field it is the duty of the informed sceptic to speak out. Unless we do so the fanatics could soon make a laughing stock of us all. Besides, in my judgement, cosmology is already taking up an unseemly share of the money and the time of the astrophysical profession – mainly for marketing reasons that are always suspect. Every one of the sixty-plus civilizations studied by anthropologists has craved for a cosmological story – and dreamed one up. Every single one. There is no need to give in to this understandable but atavistic craving today.

References

[1] Disney M. J., 2000, *General Relativity and Gravitation*, **32**, p. 1125, *astro-ph 0009020*
[2] Boorstin, Daniel, 1985, *The Discoverers*, Vintage Books, p. 86

Panel discussion

Francesco Bertola

Dipartimento di Astronomia, Università di Padova, Italy

In 1976, twenty eight years ago, Jean-Claude Pecker was the promoter of a meeting in Paris titled, "Décalages vers le rouge et expansion de l'univers; l'évolution des galaxies et ses implications cosmologiques." The meeting had to be split into two parts, one under the auspices of the International Astronomical Union, and the other sponsored by the Centre National de la Recherche Scientifique, which seemed more inclined to allow the expression of less orthodox views. In fact, several researchers presented papers mainly on anomalous redshifts and on the possibility they were not of cosmological origin.

Under the impulse of Jean-Claude Pecker and Jayant Narlikar the present meeting has been dealing with facts and problems in cosmology, with some emphasis on alternative cosmologies. A quite natural question follows: What are the achievements of the last twenty eight years in favor of the two positions in cosmology, the orthodox and the heterodox one? There is no doubt that the last decades marked a triumph of classical cosmology. The study of dark matter in galaxies and in clusters, the structure of the background radiation, the determination of the Hubble constant, the study of distant supernovae inferring the presence of the dark energy, and the understanding of the large-scale structures we heard about during this meeting mark the great success of so-called orthodox cosmology.

Disturbing evidence was presented in the nice paper given by Margaret Burbidge, where several interesting and unexplained cases of association of quasars and low-z galaxies were presented. These facts are not easily interpretable in classical terms, unless a kind of lensing effect is active. Puzzling is the case of the association of a QSO with the galaxy NGC 7319 presented by Halton Arp. These phenomena raise the fundamental problem of the nature of the redshift, for which, however, no satisfactory explanation exists except the cosmological one. In addition, one should note that there is no wide acceptance of these kinds of results among the astronomical community.

Thirty years ago, phenomena of fission of galaxies (and consequently the ejection of quasars from galaxies) were popular under the influence of the ideas of V. A. Ambartsumian. Nowadays it is believed instead that merging phenomena are the rule.

In conclusion, while in the 1950s it was possible to speak of rival theories in cosmology, we must conclude that now the big-bang picture has no strong rivals. This is confirmed by the fact that out of 1500 members of the IAU Division VIII (Galaxies and the Universe) only a dozen, although bright people, devote their time to the heterodox views.

21

General discussion

Q : M. MOLES :

There is a theoretical possibility to differentiate between stationary and evolutionary models: To look for cosmic evolution, in particular to verify $T_{CMBR} = T_o(1 + z)$.

A : G. BURBIDGE :

You are correct, and attempts are being made to verify that $T_{CMBR} = T_o(1 + z)$.

A : A. BLANCHARD :

Yes, this is a way to test expansion. It has already been attempted and results are consistent with the standard picture. The time dilatation of the apparent duration of SNIa light curves is another interesting test whose results also agree with the standard picture.

Q : M. MOLES :

It has been said that we cannot ask a theory to integrate all the observational facts at once. What we would need, if the aim is to build an alternative cosmology, is a change in perspective.

Both the standard and quasi-static cosmology accept expansion as the primary mechanism to understand the z-phenomenon. Whereas it is perfectly acceptable, it rests on the hypothesis that the general behavior of the space-time is the cause for the observed z-distance relation. A completely different view can be put forward, trying to look for a different explanation for the z-phenomenon. This could then be, in principle, tested at the laboratory level, as stated by Zwicky in 1929. In those views, started by Pecker, Vigier, Molès, and others, the CMBR could be understood as a phenomenon related to the z-phenomenon.

A : G. BURBIDGE :
Both standard cosmology and the QSSC accept expansion as the primary mechanism to understand the z-phenomenon. We do this because it naturally explains the redshift in terms that are acceptable to known physics. For normal galaxies of stars it works well except for a very small periodic term first discussed by Tifft. The "tired light" hypothesis originally proposed by McMillan and Zwicky and revived by Born is simply incompatible with experimental measurements in atomic physics, because, accompanying any loss of photon energy, there will be scattering, which is clearly not present.

Q : R. KEYS :
In the 1917 paper where Einstein did introduce the cosmological constant the motivation was given as the need to produce a quasi-static distribution of matter, as known stars appeared to possess no large-scale, relatively high velocities.

A : A. B. :
I think the "static state of the Universe" condition was not central to Einstein's thoughts (rather an obvious a priori). However, much more important to his eyes was to avoid some divergence at infinity, like in the Newtonian potential (which he claimed would produce large velocities).

A : G. BURBIDGE :
Einstein lived in an era in which astronomers only knew about the Milky Way. This was the whole Universe according to them and *it is static*.

    Einstein, could only get agreement with a static model if he put in a non-zero cosmological constant, so this is what he did.

A : A. BLANCHARD :
I think the "static state of the Universe" condition was not central in Einstein thought (rather an obvious a priori). However, much more important to his eyes was to avoid some divergence at infinity, like in the Newtonian potential (which he claimed would produce large velocities).

Q : D. ROSCOE :
A central theme of this meeting has been the vexing nature of redshift phenomenology. GR treats a photon as a particle purely by considering its trajectory along a null geodesic but completely ignores its nature as a manifestation of electromagnetism! My gut feeling is that we will only come to understand redshift phenomenology in its entirety when finally electromagnetism and gravitation are properly brought together.

A : A. B. :

In GR null geodesics are corresponding to rays in classical optics. One can have an electromagnetic description when needed and the redshift phenomenon is not problematic in both approaches.

A : G. BURBIDGE :

You may be correct.

Q : G. PATUREL :

I would ask if the cosmological constant (Λ) can be considered as a constant of integration. I had the feeling that it was so because it leads to a more general solution. In this sense Λ is compulsory. But of course Λ could be equal to zero, depending on external conditions (e.g., observations).

A : A. B. :

In some mathematical formulations of GR, Λ could be seen like this. This is, however, of little physical significance. In the modern view Λ appears as some special matter ("quintessence"), a scalar field, which can behave like the historical cosmological constant. In any case, one would like an origin to explain its numerical value. This is challenging because the energy associated to the "observed value" (if not zero . . .) is very low compared to the "natural scale."

A : G. BURBIDGE :

Λ is a constant of integration. But its actual value can only be determined when we compare the model and the observed Universe. The observed Universe, when interpreted in terms of this standard model, does indeed require a non-zero value for Λ. At the same time, the standard model may well be wrong.

Q : M. CASSÉ :

Can we get an acceleration of the expansion of the Universe without a cosmological constant, for instance via brane cosmology?

A : A. B. :

Under the standard general relativity (GR) an acceleration phase requires an equation of state with $p < -p/3$ (with $p > 0$). However, within an extension of GR there might be an acceleration without such a "strange matter."

A : G. BURBIDGE :

Acceleration of the expanding Universe means in general terms that as the Universe expands energy is added. That energy overcomes the tendency of the Universe to slow down owing to gravitational attraction. In the standard model this has led to "dark energy." In the quasi-steady state model it is simply due to creation in galaxy centers.

Q : A. BLANCHARD :

The absence of detection of "non-baryonic" dark matter even with future high efficiency has little impact on the actual significance of its inclusion in the model. Supersymmetric dark matter candidates are one family, but there is no indication that this is the right thing to look for.

A : G. BURBIDGE :

Of course you are correct in your belief that failure to detect any particle, which can be construed to mean that some kind of "non-baryonic" matter exists, is not an argument against the belief that non-baryonic dark matter plays a major role in cosmology. However, this means that the whole of big-bang cosmology rests on a theoretical concept for which there is no independent evidence at all. Without it you cannot form galaxies, and so the big-bang cosmology fails. After all, the existence of galaxies is, to me, at least as important as the existence of the CMB. Thus for me, the alternative theory involving a cyclic Universe (in which galaxies beget galaxies, and only baryonic matter is needed, and the CMB and abundances can be understood) is preferable. All aspects of this alternative cosmology have observational consequences and many of them are known to exist.

H. BROBERG :

The purpose of this comment is to compare predictions of gravitational redshifts with observation, using the gravitational theory developed by H. Broberg (in : *Inertia, Gravitation and Mach's principle*, 2003, Apeiron publ., Montreal). The theory is based on an extension of Einstein's Special Theory of Relativity from linear velocities, also to include acceleration, which facilitates a relatively easy link to gravitation in basic agreement with results of General Relativity. This theory, if proven to be correct, could open up a new cosmology as a viable alternative to the "big-bang" theory.

The table overleaf is made to compare quasar redshifts described by the Karlsson periodicity, well confirmed by observed data, according to G. Burbidge and W. Napier (*Astron. J.*, Vol. **121**, 21, 2001), with those derived from our theory. For details on this theory, we refer the reader to the above quoted paper.

Let us define the parameter $N = R_i/R_g$ as the ratio of the distance to the gravitational singularity in the gravitationally contracted space-time to the Schwarzschild radius. From the theory, we get N as a function of the "gravitational z": $N = (z+1)/z(z+2)$.

Table 21.1. *Comparison between predicted and measured*
quasar redshifts

$N = 16/2$	predicted $z = 0.064$	Karlsson's number 0.062
4/2	0.281	0.30
1	0.618	0.60
11/16	0.964	0.96
8/16	1.414	1.41
6/16	2.000	1.96
5/16	2.49	2.63
4/16	3.236	3.45
3/16	4.514	4.47

In the first column of the table, the redshift z has been calculated as a function of the parameter N, which is chosen to be either a natural number or a fraction of such. The second column gives predicted redshifts and the third column the closest Karlsson number, which correspond rather well to the observed peaks in the actual z distribution.

The agreement between prediction and observation supports both the conclusion that the redshifts are quantized, and the validity of the "gravitational theory" used for the predictions.

A : J.-C. PECKER :
Truly enough, this theory has indeed not be discussed during the meeting, and I therefore do not feel at ease to comment upon it. As the co-organizer of this meeting, I would like to state that a difficulty seems to me that you predict many more peaks in the z-distribution of QSOs redshifts (I reproduced, in this edited version of your original contribution, only those coinciding with Karlsson's numbers), but that only a few are really observed, which casts a doubt on the significance of this particular prediction of your model.

A : A. BLANCHARD :
Certainly. It would be interesting to have an alternative theory "ab initio" that can be tested against observations. However, from what I know, this is not the case. We are rather confronted with a situation in which some observations are considered as not fitting well into the standard model, requiring the introduction of "new physics" (although as I mentioned elsewhere, even the existence of non-standard redshift for QSOs, as suggested by G. Burbidge, might not necessarily request a revision of the standard cosmological model!). The question is when

do you consider that existing observations really require the introduction of something "exotic." For instance, introduction of dark matter, non-baryonic dark matter, and now dark energy are "exotic" ingredients introduced in order to reproduce existing observations. These ingredients are regarded as acceptable by most of the cosmologists, given the consistency of the global picture.

Q : M. MOLES :

About Λ: I would like to recall that it is a necessary ingredient of the most general theory, and there are no "a-priori" reasons to put $\Lambda = 0$.

A : A. BLANCHARD :

I am not sure I agree with you. From a specific mathematical point of view you might be right, but from the physical point of view, I do not agree completely: At the time of Einstein, the request to match Newton laws did not require the introduction of a non-zero lambda. From a modern point of view, we can always introduce arbitrary fields with unknown equations of state, in which case lambda is nothing specific.

Q : M. MOLES :

About dark matter, it is needed by the standard model to have compatibility with CMBR fluctuations (provided it is non-baryonic and cold). My question is: Would some of the panel members comment on the necessity to fine tune the amplitude of the fluctuations emerging from the inflationary epoch?

A : A. BLANCHARD (About dark matter) :

The need for dark matter is first coming from observations: Mass estimates of galaxies and clusters indicate more mass than what we actually see. This is unavailable unless one modifies gravity on "short scales" (few kpc). The need for non-baryonic dark matter is from consistency with primordial nucleosynthesis. Therefore the shape of CMBR fluctuations was predicted before being observed, and the observed angular spectrum is consistent with the shape predicted. The amplitude is therefore observed, matching precisely the amplitude predicted (which was normalized from information at $z \sim 0$). Inflation is one possible explanation of the origin of these fluctuations (and actually the only one we have . . .). Within inflation, this amplitude (10^{-5} is not regarded as very natural and is said to require some fine tuning. This is, however, a problem of the theory of inflation, which might happen to be wrong (!) without any trouble for the standard cosmological picture. It may also reflect our limited understanding of the relevant physics, a more interesting perspective.

Comment: C. GUTERRIEZ :
I believe that we want to know whether objects with high redshift are roughly at distances indicated by their redshifts. It is extremely difficult to find a standard candle, and galaxies are NOT standard candles (even at low redshift), despite that we can check that they follow the Hubble law between distance (and NOT luminosity) redshift from the existence of standard candles (like cepheids). For quasars, we just do not have reliable standard candles. These are just observational facts that are independent of any interpretation. Coming back to quasars, I believe that we have few cases that show that they actually are at the cosmological distance: For instance, the gravitational lensing was **predicted** before being observed, that multiple images of the same would be found around some galaxies, and that the time variation of the object will be identical to some difference of the order of one year.

And this has been observed! The fact that identical light curves are observed in QSOs associations when the standard theory predicts it and never when it predicts that this should not be is to me extremely convincing (again we want to have theories that make predictions that can be tested, and not "post-diction"). Similarly, the existence of galaxies associated with the absorption lines seen in quasars (with identical redshift) is also a direct indication that quasars are in the background of these galaxies. As we are now detecting galaxies with high redshift (greater than 5!), I am inclined to say that we are close to having demonstrated that quasars have as high a redshift as the galaxies.

Q : J.-C. PECKER :
What is your feeling about the polyedric topology suggested by Luminet *et al.*?

A : A. BLANCHARD :
I think the possibility of non-trivial topology that Luminet has pushed for several years is a very bright idea. It is possible to find signatures of such non-trivial topology on CMB maps so we can test such an idea. I am personally reluctant, however, to support the idea, because it needs the addition of a spatial scale, which by some coincidence has just to be close to the horizon now. (Why the spatial distance associated to topology is not just one kilometre has to be explained!)

Comment : H. ARP (To DISNEY) :
The argument that quasars must be at their redshift distance because their host galaxies are at their redshift distance is now no longer used because it is real-ized that, as a quasar expands and evolves into a galaxy, all the material in that young galaxy has the same redshift. Observations have increasingly shown the whole galaxy of companions and young galaxies have intrinsic redshifts. As for the alignments of high redshift objects across low redshift, ejecting galaxies having

a-posteriori improbabilities of being chance: The first alignment is a posteriori but every subsequent example that has the same closeness of bright objects, alignment, centering, and similarity of ejecta is an a-priori prediction, and the improbabilities compound to a ridiculously small number.

Finally, Bayesian statistics tells us that people's interpretation of statistics is influenced by their prejudices. That is not news.

Comment : F. BOUCHET : *(This remark, made during Session I does seem more appropriately located within the General Discussion. Editors)*
As pointed out by introductory speakers, there are hypotheses underlying predictions, e.g., photon to baryon ratio, existence of Dark Matter . . . to account for abundances, structures given the homogeneity of the microwave background. So what? The question is not to see the power of the theory (number of explained facts vs. number of hypotheses) but how predictive it is, etc . . .

Comment : M. DISNEY : *(This remark, made earlier during the meeting, does seem more appropriately located during the General Discussion. Editors)*
It is worth commenting that in the early days (1930s) of the redshift debate ("is it expansion?"), it was said the Tolman test would decide, i.e., surface brightness proportional to $(1 + z)^{-4}$. Well, it doesn't work! Conveniently for the conventional hypotheses, the rate of evolution of star formation is sure to just balance the $(1 + z)^{-4}$ effect.

Comment : MARTIN LOPEZ-CORREDEIRA :
(Several invited participants could unfortunately not attend the meeting. This is the case of Dr Lopez-Corredeira, from the Astronomisches Institut der Universität Basel. Dr Lopez-Corredeira has nevertheless sent a reprint of his paper, which has been distributed to participants. It is impossible to give here the totality of this paper, published elsewhere (Recent Res. Devel. Astron. & Astrophys., 1 (2003): 561–589 ISBN: 271-0002-1). We give hereafter its summary. But we should add that this paper gives a very comprehensive review, and contains an extensive bibliography of 259 entries. Editors)

Observational cosmology: Caveats and open questions in the standard model.

Abstract:
I will review some results of observational cosmology, which critically cast doubts upon the foundations of the standard cosmology: (1) The redshifts of the galaxies

are due to the expansion of the Universe. (2) The cosmic microwave background radiation and its anisotropies come from the high energy primordial Universe. (3) The abundance pattern of the light elements is to be explained in terms of the primeval nucleosynthesis. (4) The formation and evolution of galaxies can only be explained in terms of gravitation in the cold dark matter theory of the expanding Universe. The review does not pretend to argue against this standard scenario in favor of an alternative theory, but to claim that cosmology is still a very young science and should leave the door wide open to other positions.

Comment : J.-C. PECKER :

I would like to make three comments on points that have not been mentioned so far in the discussion of the panel.

A. One has questioned the restriction of the cosmological debate to Einstein's General Relativity (GR).

I see no compulsive reason to move from GR to something better, i.e., more general. But I strongly challenge the use (by Einstein, notably) of the so-called "cosmological principle," more a simplicative assumption than a real "principle," introduced indeed in order only to solve easily the GR tensorial equations, by limiting the solution to the computation of one quantity only, the "scale factor," and to needing a minimum number of integration constants. Not only the Universe, as we see it within the limits of our observations, is neither uniform nor isotropic, but, moreover, at least between two scales, the density distribution is clearly fractal (as shown by de Vaucouleurs, years ago).

But we do not know how to properly solve the GR equations in such conditions; we cannot even assume that if the usual assumptions are valid at very large scales, the equations can be solved without considering what happens at smaller scales, which actually put some restrictions upon the boundary conditions.

B. But of course, one can also consider that something is missing in GR, although I am personally reluctant to do so. There have been several attempts along this line, in order not to abandon it but to extend the GR theory.

(i) The Brans–Dicke theory, of which the GR is a particular case, is an important proposal to remember.

(ii) MOND (modified Newtonian dynamics) is, as explained to us by David Roscoe, a very suggestive "recipe"; but it may reveal some deeper meaning. We should not forget, however, similar attempts made in the 1890s, by von Seeliger, notably, to multiply the Newtonian term in $1/r^2$ with an exponential term e^{-kr}, in order to solve the problem of the advance of the Mercury perihelion.

(iii) Segal's chronogeometry has not been really developed to its ultimate consequences; but some observations (Segal & Nicholl, in the 1970s for example) give some weight to it, and should be re-examined with care.

(iv) There are GR cosmologies, making an extensive use of exotic topologies, such as the polyhedric Universe of J.-P. Luminet, or the twin Universes system of J.-P. Petit, or again the Wheeler double sheets.

(v) Etc . . .! No cosmologic theory should be considered a priori to be eliminated, just because some others seem to have accounted for a larger field of observations. In no case, as very wisely noted by Narlikar (2004), does cosmology have the possibility of being checked properly; its processes can not be reproduced. It is neither verifiable, nor indeed falsifiable. Therefore, one should be more tolerant of cosmologies often qualified as exotic.

C. I would like to make a third comment. I was asked by Henrik Broberg (who was obliged to leave the colloquium before this morning to return to Sweden), to mention his recent paper edited by Sacks and Roy (Apeiron 2003). It describes a theory of interactions between matter and vacuum energy-fields, incorporating basically the Mach's principle as its main ingredient. That kind of physical theory should be taken very seriously. See his comment hereabove, in the present discussion.

Comment : F. M. SANCHEZ[1] :

Towards the Grand Unified "holic[2]" Theory. Henri Poincaré had predicted in 1912 the existence of a unified cosmical theory; but as *"the Universe exists in only one copy,"* this theory would be devoid of any free parameter. *"Its laws could not be expressed by differential equations."* In other terms, it means a sort of general quantification, in an absolute cosmical reference framework (ACR), made concrete by the observation of the background radiation of temperature $T_{ph} \approx 2.725(1)$ K, and by the cosmical oscillations (without Doppler effect) of period $t_K \approx 9600.6(1)$ s. But this second fact, although much confirmed during the last 30 years (Kotov *et al.* 2003), has been unduly rejected ever since.

However the "micromachian" non-local radius (independent of c) given by elementary dimensional analysis (EDA) is $R = \hbar^2/Gm^3$ – the famous Eddington's formula, which gives, together with Nambu's mass $m_{Nb} \equiv \alpha^{-1}m_e$, a radius close to the "Hubble radius": $R_\alpha \approx 2R_H/3$. By a symmetrization of the three "bricks" of the Universe (namely: proton, neutron, electron), one gets: $\hbar^2/Gm_p m_n m_e \approx R_H/2$ (Sanchez, Sept. 1997), or $R_H \approx 13.8 \times 10^9$ years, the so-called "Hubble age" being 13.7 ($\pm$ 0.2) 10^9 years. The "critical relation" (Schwarzschild): $R/2 = GM/c^2$ (of a type

---

[1] *(This contribution has been translated from the French by the editors; it appeared very difficult to present it in a more condensed version. Therefore, the editors decline all responsibility w.r. to the content of this paper. Editors).*

[2] In this paper, the prefix *holo-*, the words *holic* or *holistic*, point out the fact that the author's ideas are in a large part inspired by the physics of holography.

"macromachian EDA"), then unifies the well-known double correlation between "large numbers" $R_H/2\lambda_p \approx \hbar c/Gm_e m_p \approx (M_H/m_e)^{1/2}$, and symmetrizes the masses: $M_H m_e m_p m_n \approx m_{Pl}^4$. Moreover, $2\hbar^2/Gm_p^2 m_e$ is the limit value of a stellar radius, for a single couple electron-proton, this confirming the "superquantic cosmic scanning" (Sanchez, 1995), which unifies indescernability, exclusion, and non-locality: Only one particle per species is involved in matter–antimatter oscillations. Replacing then $Gm_p m_n$ by $e^2 \equiv \hbar c/137,036$ allows to find the Bohr's radius r_B, also predicted by EDA: The *Universe can be considered as a giant atom*. The logarithmic electro-gravitational interaction (2), is well illustrated in the model of the single spiraling electron, "spiraling" till R_H, giving again the value of r_B and, by iteration, $l_K = ct_K \approx$ Sun–Uranus distance, the key for understanding the Solar System, according to Kotov and related by $R_H l_K^2 \approx l_L^3$ to the length l_L to the solar cycle, i.e., $c \times (10^9$ years).

The "holic principle" (Sanchez 1994) assumes the universal conservation of information in the numerical quest of a calculatory diophantian cosmos. It uses the physical parameters as the best bases of any computation. And $R_H/2\lambda_e$ is very close to 2 at the power $2^{6.9999}-1$, – the last term of the "combinatory hierarchy (CH)" of ANPA (Sanchez 1994), summing $x' = 2^x - 1$, and of which two terms[3] are 10 and 137, closer to the inverse coupling constant of nuclear weak and electromagnetic interactions. Relations derived from macrophysics (Sanchez, 1987–1998) demonstrates the existence of a hidden natural mathematics, of the "holic" type, and including a diophantian degeneracy. The following "equivalences": $n_{ph}/(M_H/m_e) \approx m_e c^2/E_{Ryd} \equiv 2\alpha^{-2}$ and $M_H c^2/E_{ph} \approx \alpha^{-2}$, lead to $T \approx 2.67$ and 2.78, both sides of $T_{ph} \cong 2.725$ K. The partial success of the big-bang cosmology, here radically disproved, could be due to the tunnel effect, during the long durations implied by the stationary Universe of Hoyle, Bondi, & Gold, where the accelerated expansion is at once justified by entropic considerations. Now, one has: $M_H c^2/E_{ph} \approx E_{Ryd}/4kT_{neut}$ (0.1%); $n_{ph}/(M_H/2m_p) \approx (M_H c^2/E_{ph+neut})^2$ (1%); and $M_H E_{neut}/E_{ph} \approx m_e n_{ph} kT_{ph}/E_{Ryd}$ (1%): The three cosmic neutrinos would indeed exist (the missing mass?). "Holic" equations are generated by the transdimensional geometrical conservation of a "number of information channels," which would be the machian concept unifying the "quantinuum"[4] Space–Matter: this is the "Holographic principle (HP)" (Sanchez, 1992). Such is the micromachian theory, with $m_p = m_n$, of which the prolongation towards volumes leads to $\lambda_{ph} \equiv \hbar c/kT_{ph}$, in common with an ACR holography $(l_K, \lambda_p, \lambda_e)$, of which the elimination gives $R_H/l_K \approx 4(m_p/m_e)^4$.

---

[3] The prime number 137 is the famous hidden "monster" of the Egyptian progression $1/n$ (for $n = 5$) and could well be known by the Egyptians (Sanchez, 1998): the Amon's temple, at Karnak, illustrates 137 and the CH by its 134 columns, ranked by 7 double columns of the hypostyle room, plus the 3 terminal pylons.

[4] The need for the quantization of space-time led us to introduce that clear neologism.

The critical condition is defined by the universal holographic principle (U-HP)(Sanchez, 1998): $N_{\text{channels}} \equiv 2\pi R_H/\lambda_M = \pi \, (R_H/l_{\text{Pl}})^2$, where the linear quantum is the reduced Hubble's wavelength $\lambda_M \equiv \hbar/M_H c$ (the "topon" $\approx 4 \, 10^{-96}$ m), and the surface quantum is the Plancks's area λ_{Pl}^2. One thus finds directly the "black-hole entropy" of Bekeinstein–Hawking. The critical character of our computations expresses the fact that one could consider the Universe as a black hole. Hawking has recently accepted that the entropy of a black hole is conserved, against the opinion of t'Hooft, although the latter is considered as the discoverer of HP (1993). The nucleon's radius $r_{\text{cl}} \equiv a\lambda_e$ then appears as the volume-quantum in an open geometry: $N_{\text{channels}} \approx (2\pi/3) \, (R_H/r_{\text{cl}})^3$; and λ_K as the geometrical closure radius characterizing the 3D topology: $\lambda_K^3 = r_{\text{cl}} R_H^2/2$. As the sphere can be generated by the transverse scanning of a great circle, to any particle of mass m is associated a number of Universe loops M_H/m, – a cosmic definition of wavelength, scaled by the topon. The "strings" would not be folded but unfurled. Actually, for the electron, the U-HP can be written using $\exp(2^{n/4})$, which makes appear the 4 gauge bosons, thus justifying $m_X/m_e \approx (m_W/m_e)^4$ (Carr and Rees, 1979), and the special dimensions $n = 2 + 4p$ of strings. This "Topological axis" (Sanchez, May 1998) suggests that the Universe is a gauge boson in the "grand cosmos" of radius $R_{\text{GC}} \equiv R_a^2/2l_{\text{Pl}}$, where the relative energy of vacuum (of which the limit is r_B) is equal to $(4\pi^2/3)(R_{\text{GC}}/r_B)^3 \approx 137{,}036^{137{,}036(1)}$: α^{-1} is indeed an optimum basis of our construction.

The ADE statistically confirms the invariance of all used physical quantities and attributes a central role to $n\lambda_{\text{ph}}$, close to $\lambda_{\text{Ryd}}^2\lambda_e$, and to the mass of the codon: $m_{\text{cd}} \approx m_p^2/m_e \approx 1836\,m_p \approx M_H/\,2^{2\times127}$ in $m_{\text{cd}}^2/kT_{\text{ph}} \approx m_{\text{Pl}}/c^2 \equiv l_{\text{Pl}}/G$, that of the average nucleotide being $m_{\text{cd}}/6 \approx m_{\text{Fermi}} \approx m_p l_K l_{\text{Pl}}/\lambda_e^2$, emphasizing the "geometrical physics" of Wyler: $1836 \approx 6\pi^5$: the DNA would be a helix hologram! Moreover, $T_{\text{PHU}} \equiv hc/k(R_H l_{\text{Pl}})^{1/2} \approx jT_{\text{ph}}$ soit 37.3 °C, with $j \approx 8\pi^2/\ln2$, the Sternheimer's scaling factor. For R_α, the U-HP is associated to the power 12 of $T_{\text{ph}} \cong 2.728$ K, whenever $T_{\alpha\text{U-HP}}$ is nearly equal to 273 K, coming from the triple points relation $T_{\text{hyd}}T_{\text{oxy}} \approx T_{\text{water}}T_{\text{ph}}$. Biology, as a management of information at a key temperature (Schrödinger), is indeed included in the Great Holic theory that we annnounce. Unlike the anthropic principle (of Dicke and of Carter), refuted together with the big-bang cosmology, by a simple use of EDA, it gives the role of Life: to serve the cosmic quest. Thus the musical sense would be the multibasis unconscious calculus. We are back to the Pythagorian *"everything is number."*

References

Carr B. J. and Rees M., 1979, The anthropic principle and the structure of the physical world, *Nature*, **279**.

Hawking S., 2004, *New Black Hole Theory. General Relativity Gravitation 17*. Dublin, July 2004.

Kotov V. *et al.*, 2003, *Izv. Krym. Astrofiz. Obs.* Vol. **101** (2003).

Sanchez F. M. The author, founder of the French school of holography, has submitted his discoveries to several journals. They all have been refused by anonymous reports, even after the observation of the predicted acceleration of expansion, in 1999. His model has, however, been tested in microphysics since 1987; he presented it to the Fondation Louis De Broglie, then to ANPA (Cambridge, Sept. 1994), which published the *Holic Principle* in 1995.

22

Concluding remarks

Jayant V. Narlikar

College de France, Paris and Inter-University Centre for Astronomy and Astrophysics, Pune, India

I have a dual role to play: As convener of this Panel Discussion and as a co-organizer of this meeting. So what I have to say is a mixture of the two.

First I wish to thank the panelists for expressing their points of view succinctly and also replying to the comments from the floor. Their differing points of view are what this meeting is about, namely a free and frank discussion of our current ideas about the origin of the Universe.

The majority view is expressed by Bertola and Blanchard, while the skeptical minority view is expressed by Disney and Burbidge. The majority view is simply that the standard model with specified parameters, having rather precisely determined values, explains all known facts of the Universe and that there is now a consensus amongst the community that this is so. Whatever remains to be understood can and should be explained only within this framework. This view is today called "precision cosmology" or "concordance cosmology." Besides stating this premise, the view is that no other way of understanding the Universe has gone down to the same level of detail as the standard model and therefore all such alternatives cannot be compared to the standard model.

The minority view is that the successes claimed by the concordance model have been achieved at the expense of certain assumptions that have not been independently tested. These include inflation, non-baryonic dark matter, dark energy, etc. and the very high energy physics used as the basis of these ideas has not been tested in the laboratory. Further, there remain observed peculiarities, especially about redshifts, that cannot be understood by the standard model. These were discussed in this conference. Given the epicyclic nature of the standard model and its inability to understand these phenomena, it is wiser to keep an open mind about cosmology and also, attempt alternative ideas provided they are worked out in full detail.

Perhaps I should add here the fact that one reason that the alternative ideas are not worked out in as much detail as concordance cosmology is because there are so few people working on them. There is a vicious circle here . . . by discouraging

261

young people to take on research in a non-big-bang cosmology on the grounds that it is a fruitless exercise, one is severely limiting the development of alternatives. Research grants and project funds, and observing time on telescopes are hard to come by if they aim to investigate any alternative model.

But is the concordance model really close to understanding the Universe? Perhaps I should quote here a comment made by the late Fred Hoyle at the Vatican Conference of 1970:

> ... *I think it is very unlikely that a creature evolving on this planet, the human being, is likely to possess a brain that is fully capable of understanding physics in its totality. I think this is inherently improbable in the first place, but even if it should be so, it is surely wildly improbable that this situation should just have been reached in the year 1970* ...

Hoyle's cautionary remarks were directed against those enthusiastic proponents who were categorically pushing the then "accepted" version of the big-bang model. In hindsight we can see the caution justified even in the framework of today's standard cosmologists. The 1970 model had a decelerating Universe, no lambda term, no non-baryonic dark matter, no inflation, no high-energy physics. I personally feel that even today we should remember those cautionary words. As the observing tools of cosmologists are improving, new aspects of the Universe are coming to light. To say that everything about the Universe is known today is like the nineteenth-century physicist claiming to have reached the end of physics two decades before the advent of special relativity and quantum theory. In those days discrete spectral lines were not understood. Today we may wonder in the light of the evidence presented at this meeting whether we really understand the redshift.

Having said that I also feel that there is a need to explore alternatives to greater depth. I wish the climate of research was benign enough to encourage new ideas in cosmology. I recall Subrahmanian Chandrasekhar relating the episode in which Hubble and Eddington were being interviewed for their comments on the news that the 5-metre telescope had been approved for construction on the Palomar Mountain. When the reporter asked: "What do you expect to find with this new experiment?," their reply was: "If we knew the answer, we would not have proposed such an instrument." Contrast this 1930s episode with any in modern times. If a telescope like WMAP is proposed, the proposers must state in detail what they expect to find with it. If the proposal is consistent with the standard paradigm, it would be accepted, not otherwise. This is the reason why alternatives have not flourished to the level they should.

Which is why a meeting of this kind is very much needed, a meeting in which the different points of view confront one another and much is learnt about the other's point of view. I am indeed grateful to Collège de France for holding this meeting and the Foundation Hugot for the support it has extended.

But most of all I thank my friend and colleague Jean-Claude Pecker for taking the initiative to organize this meeting during my tenure here as Professor in the Chaire Internationale. It has been he, almost single-handedly, who has run this meeting and helped on various occasions with organizational problems.

Finally I thank you the participants for making the meeting an interactive one as we had planned.

Index

Printed in the United States
By Bookmasters